DOSIMETRY IN BRACHYTHERAPY – AN INTERNATIONAL CODE OF PRACTICE FOR SECONDARY STANDARDS DOSIMETRY LABORATORIES AND HOSPITALS

The following States are Members of the International Atomic Energy Agency:

AFGHANISTAN	GAMBIA	NORWAY
ALBANIA	GEORGIA	OMAN
ALGERIA	GERMANY	PAKISTAN
ANGOLA	GHANA	PALAU
ANTIGUA AND BARBUDA	GREECE	PANAMA
ARGENTINA	GRENADA	PAPUA NEW GUINEA
ARMENIA	GUATEMALA	PARAGUAY
AUSTRALIA	GUINEA	PERU
AUSTRIA	GUYANA	PHILIPPINES
AZERBAIJAN	HAITI	POLAND
BAHAMAS	HOLY SEE	PORTUGAL
BAHRAIN	HONDURAS	QATAR
BANGLADESH	HUNGARY	REPUBLIC OF MOLDOVA
BARBADOS	ICELAND	ROMANIA
BELARUS	INDIA	RUSSIAN FEDERATION
BELGIUM	INDONESIA	RWANDA
BELIZE	IRAN, ISLAMIC REPUBLIC OF	SAINT KITTS AND NEVIS
BENIN	IRAQ	SAINT LUCIA
BOLIVIA, PLURINATIONAL	IRELAND	SAINT VINCENT AND
STATE OF	ISRAEL	THE GRENADINES
BOSNIA AND HERZEGOVINA	ITALY	SAMOA
BOTSWANA	JAMAICA	SAN MARINO
BRAZIL	JAPAN	SAUDI ARABIA
BRUNEI DARUSSALAM	JORDAN	SENEGAL
BULGARIA	KAZAKHSTAN	SERBIA
BURKINA FASO	KENYA	SEYCHELLES
BURUNDI	KOREA, REPUBLIC OF	SIERRA LEONE
CABO VERDE	KUWAIT	SINGAPORE
CAMBODIA	KYRGYZSTAN	SLOVAKIA
CAMEROON	LAO PEOPLE'S DEMOCRATIC	SLOVENIA
CANADA	REPUBLIC	SOUTH AFRICA
CENTRAL AFRICAN	LATVIA	SPAIN
REPUBLIC	LEBANON	SRI LANKA
CHAD	LESOTHO	SUDAN
CHILE	LIBERIA	SWEDEN
CHINA	LIBYA	SWITZERLAND
COLOMBIA	LIECHTENSTEIN	SYRIAN ARAB REPUBLIC
COMOROS	LITHUANIA	TAJIKISTAN
CONGO	LUXEMBOURG	THAILAND
COSTA RICA	MADAGASCAR	TOGO
CÔTE D'IVOIRE	MALAWI	TONGA
CROATIA	MALAYSIA	TRINIDAD AND TOBAGO
CUBA	MALI	TUNISIA
CYPRUS	MALTA	TÜRKİYE
CZECH REPUBLIC	MARSHALL ISLANDS	TURKMENISTAN
DEMOCRATIC REPUBLIC	MAURITANIA	UGANDA
OF THE CONGO	MAURITIUS	UKRAINE
DENMARK	MEXICO	UNITED ARAB EMIRATES
DJIBOUTI	MONACO	UNITED KINGDOM OF
DOMINICA	MONGOLIA	GREAT BRITAIN AND
DOMINICAN REPUBLIC	MONTENEGRO	NORTHERN IRELAND
ECUADOR	MOROCCO	UNITED REPUBLIC OF TANZANIA
EGYPT	MOZAMBIQUE	UNITED STATES OF AMERICA
EL SALVADOR	MYANMAR	URUGUAY
ERITREA	NAMIBIA	UZBEKISTAN
ESTONIA	NEPAL	VANUATU
ESWATINI	NETHERLANDS	VENEZUELA, BOLIVARIAN
ETHIOPIA	NEW ZEALAND	REPUBLIC OF
FIJI	NICARAGUA	VIET NAM
FINLAND	NIGER	YEMEN
FRANCE	NIGERIA	ZAMBIA
GABON	NORTH MACEDONIA	ZIMBABWE

The Agency's Statute was approved on 23 October 1956 by the Conference on the Statute of the IAEA held at United Nations Headquarters, New York; it entered into force on 29 July 1957. The Headquarters of the Agency are situated in Vienna. Its principal objective is "to accelerate and enlarge the contribution of atomic energy to peace, health and prosperity throughout the world".

TECHNICAL REPORTS SERIES No. 492

DOSIMETRY IN BRACHYTHERAPY
–
AN INTERNATIONAL CODE OF PRACTICE FOR SECONDARY STANDARDS DOSIMETRY LABORATORIES AND HOSPITALS

INTERNATIONAL ATOMIC ENERGY AGENCY
VIENNA, 2023

COPYRIGHT NOTICE

© IAEA, 2023

Printed by the IAEA in Austria
December 2023
STI/DOC/010/492

IAEA Library Cataloguing in Publication Data

Names: International Atomic Energy Agency.
Title: Dosimetry in brachytherapy – an international code of practice for secondary standards dosimetry laboratories and hospitals / International Atomic Energy Agency.
Description: Vienna : International Atomic Energy Agency, 2023. | Series: Technical reports series (International Atomic Energy Agency), ISSN 0074-1914 ; no. 492 | Includes bibliographical references.
Identifiers: IAEAL 23-01619 | ISBN 978-92-0-113923-8 (paperback : alk. paper) | ISBN 978-92-0-114023-4 (pdf) | ISBN 978-92-0-114123-1 (epub)
Subjects: LCSH: Radiation dosimetry. | Radioisotope brachytherapy. | Radiation — Measurement. | Radiotherapy.
Classification: UDC 615.849 | STI/DOC/010/492

FOREWORD

One of the key roles of the IAEA is to improve the traceability, accuracy, and consistency of clinical radiation dosimetry measurements in Member States. With reference to harmonization of dosimetry in external radiotherapy beams, the IAEA has disseminated a number of international codes of practice, which are published in the IAEA's Technical Report Series (TRS), providing detailed descriptions of the instruments and steps to be taken for absorbed dose determination in water.

Technical Reports Series No. 277 (second edition), Absorbed Dose Determination in Photon and Electron Beams and Technical Reports Series No. 381, The Use of Plane Parallel Ionization Chambers in High Energy Electron and Photon Beams, both published in 1997, were based on air kerma calibration standards. Technical Reports Series No. 398, Absorbed Dose Determination in External Beam Radiotherapy, which was published in 2000, was based on the application of standards of absorbed dose to water. More recently, Technical Reports Series No. 483, Dosimetry of Small Static Fields Used in External Beam Radiotherapy, was published to provide information on the dosimetry of small static photon fields used in newer techniques and technologies.

The brachytherapy process also requires consistent reference dosimetry that is traceable to metrological primary standards. IAEA-TECDOC-1274, Calibration of Photon and Beta Ray Sources Used in Brachytherapy has been a key resource for brachytherapy dosimetry since 2002. However, several new developments have taken place, in terms of available dosimetry standards, detectors, radioactive sources, and brachytherapy technologies. Following recommendations from the 17th Scientific Committee of the IAEA/WHO Network of Secondary Standards Dosimetry Laboratories (2016), it was decided to prepare an international code of practice for brachytherapy dosimetry.

This code of practice is addressed to both secondary standards dosimetry laboratories and hospitals and is based on the use of well-type re-entrant ionization chambers. It applies to all brachytherapy sources with intensities measurable by such detectors. The dosimetry formalism; common procedures for reference dosimetry and for calibration; reference-class instrument assessment; and commissioning of the well-type chamber system are described. Guidance and recommendations provided here in relation to identified good practices represent expert opinion but are not made on the basis of a consensus of all Member States.

Miniature systems that use low-energy X ray sources, usually referred to as electronic brachytherapy, are discussed in this publication. However, work is still needed at the metrological level to provide a standardized and well established approach for their dosimetry. Therefore, even if much of the content of this publication might be relevant, electronic brachytherapy sources are not included in the main section of this publication. Beta emitting ophthalmic eye plaques and applicators are also excluded from the main section. Detectors different from well-type chambers are used for their calibration. Other suitable detectors that could be used are also discussed in this publication.

The IAEA wishes to express its gratitude to all those who contributed to the drafting and review of this publication, in particular T. Bokulic (Croatia), L. A. DeWerd, (United States of America), M. McEwen (Canada), M. J. Rivard (United States of America), T. Sander (United Kingdom), T. Schneider (Germany) and P. Toroi (Finland). The IAEA also wishes to acknowledge the following people for their valuable comments and suggestions: J. T. Alvarez-Romero (Mexico), Sudhir Kumar (India), and E. Mainegra-Hing (Canada). The IAEA officer responsible for this publication was M. Carrara of the Division of Human Health.

CONTENTS

1. INTRODUCTION

1.1. BACKGROUND

Brachytherapy is a specific modality of radiation therapy in which small encapsulated radiation sources are inserted into or near the volume to be treated [1]. Historically, the term brachytherapy referred to the use of *radioactive* sources. They were in fact the only sources of radiation that could be achieved in small dimensions available at the time of brachytherapy inception, which was at the beginning of the twentieth century. More recently, miniature systems that use electronically created low energy X rays instead of radionuclides were designed [2]. At the time of writing, a few of such devices are capable of performing intracavitary or intraoperative brachytherapy treatments [3] but radionuclides remain the primary sources used.

The clinical efficacy of brachytherapy is attributable to its capability of delivering a high radiation dose to the treated volume, while limiting the absorbed dose to surrounding tissues. Brachytherapy has shown its effectiveness, especially for the treatment of specific disease sites in the body. For example, there is a high incidence of advanced cervical cancer [4] which is best treated with a combination of external beam radiotherapy (EBRT) and brachytherapy [5, 6]. Apart from cancers of the cervix uteri, major indications for brachytherapy are endometrial, breast and prostate cancer [7]. For prostate cancer treatment, for instance, high risk groups of patients treated with EBRT and boosted with brachytherapy showed significantly better outcomes than those treated with EBRT alone or undergoing radical prostatectomy [8]. Further, small tumours that are accessible for implantation can in many cases be treated with brachytherapy as monotherapy [7].

Brachytherapy is an essential modality in low and middle income countries with a high incidence of cervical or oesophageal cancer. It is also broadly disseminated in high income countries. According to the Directory of Radiotherapy Centres (DIRAC) maintained by the International Atomic Energy Agency [9], currently 3345 brachytherapy facilities are available worldwide, with 65% of these being in high income countries and 35% in low and middle income countries. Most brachytherapy procedures are now performed using high dose rate (HDR) remote afterloaders. Remote afterloading low dose rate (LDR) equipment has been discontinued by the manufacturers, leaving HDR or pulsed dose rate (PDR) brachytherapy as the major alternative technologies and restricting LDR applications to manual procedures using low energy sources.

Since HDR brachytherapy techniques deliver very high dose rates to the point of prescription (i.e. they can reach a few hundreds of Gy h^{-1} at 1 cm distance from the source) [10], mistakes can lead to a wrong dose delivery with the potential for adverse effects [11]. According to the International Commission on Radiological Protection (ICRP) [12], "more than 500 HDR accidents (including one death) have been reported along the entire chain of procedures". The involved dose rates with LDR sources are significantly lower than with HDR, but applications with such types of sources may also be the subject of misadministration, leading possibly to adverse consequences [13–15]. Even if most radiation incidents were caused by human errors, appropriate dosimetry is essential to reduce the risk of misadministration. Source strength measurement is therefore considered a fundamental part of a general quality assurance (QA) programme for brachytherapy treatments, in order to deliver the prescribed dose to the target tissues [16–19]. End user dosimetry of brachytherapy sources is necessary to ensure traceability through secondary standards dosimetry laboratories (SSDLs) to the internationally accepted standards of primary standards dosimetry laboratories (PSDLs).

The majority of HDR systems in use worldwide are ^{192}Ir radionuclide based. A very small number of these are PDR systems, which combine the advantages of HDR stepping source dosimetry principles and safety with the favourable radiobiological properties of LDR brachytherapy applications [20]. Because of the relatively short half-life and the need for regular source replacement of ^{192}Ir, other radionuclides, such as ^{60}Co [21–23], or X-ray electronic brachytherapy (eBT) devices [3, 24] have been suggested for performing HDR treatments. Other radioactive photon-emitting sources, with lower energies and dose rates than ^{192}Ir and ^{60}Co, are also widely available [25]. Each of these types of sources have their own dosimetry requirements for the PSDLs, the SSDLs and hospitals.

Beta emitting radionuclides such as ^{90}Sr/^{90}Y and ^{106}Ru/^{106}Rh are used for specialized procedures, especially concerning intravascular applications [26] (^{90}Sr/^{90}Y) or ophthalmic treatments [27, 28] (both ^{90}Sr/^{90}Y and ^{106}Ru/^{106}Rh). Beta particles generally require dosimetry at the millimetre range, whereas photon sources extend further, with an application distance that might reach up to a few centimetres in some cases. The use of surface applicators for treatments using brachytherapy sources is also growing and requires its own consideration [29]. There have been many new radioactive sources introduced, many of which have still not found a place in the community for various reasons [30–35]. In addition to the standard source strength-specifying quantity of reference air kerma rate (RAKR) or air kerma strength (AKS), the other quantity that has been suggested is absorbed dose to water. The appropriate sections below consider these in more detail.

The expansion of the use of various sources has greatly increased around the world, but some of the available brachytherapy codes of practice are on the order of 20 years old and need updating [36, 37]. The need for an international dosimetry code of practice has become evident, especially for the standardization of quantities and dosimetry procedures.

1.2. OBJECTIVES

The present International Code of Practice for Brachytherapy Dosimetry is aimed to enable common procedures to perform dosimetry of radioactive sources used in brachytherapy, excluding beta-emitting eye plaques and applicators, as well as stranded seeds and mesh type sources. Targeted radionuclide therapy and miniature X-ray brachytherapy devices, known also as electronic brachytherapy (eBT), were also excluded from this code of practice. It provides a description of the most accurate and sensitive calibration systems available at PSDLs and recommends suitable detectors and procedures for source strength measurements at SSDLs and hospitals.

Guidance and recommendations provided here in relation to identified good practices represent expert opinion but are not made on the basis of a consensus of all Member States.

1.3. SCOPE

This code of practice covers methods that are relevant to the brachytherapy dosimetry process. It is important to all the professionals involved in this process, starting from the radiation metrologist establishing the quantities at the PSDL to the clinically qualified medical physicist working in the hospital and providing the measured quantity to the treatment planning system (TPS). It addresses the main radioactive HDR and LDR brachytherapy sources currently used in the clinical practice, both photon and beta emitters, that can be measured by means of a well type re-entrant ionization chamber. This easy-to-use reliable detector was chosen as the reference detector recommended by this code of practice, since it has been used for the quantification of radioactive sources over many decades and has demonstrated its value at all levels of the calibration chain [16, 38, 39].

This code of practice is directed to the clinically qualified medical physicists and the radiation metrologists dealing with brachytherapy dosimetry and detector calibration. It is not directed to the physician involved in the clinical practice.

1.4. STRUCTURE

This code of practice consists of ten sections and six appendices. Following this introduction that frames the background and scope of this code of practice, Section 2 provides a description of the radioactive sources currently available for brachytherapy. The dosimetric quantities reference air kerma rate, air kerma strength and absorbed dose to water are discussed in Section 3, along with the dose rate constant and other parameters important to properly characterize radioactive sources. Since this code of practice is based on the use of the well-type ionization chamber instrumentation, Section 4 provides a detailed description of this instrumentation and defines the requisites for reference-class well-type ionization chambers. It also includes a description of HDR remote afterloaders. Section 5 contextualizes the dosimetry framework that defines dissemination of primary dosimetry standards down to the hospital level and Section 6 provides an overview of the available primary standards useful for brachytherapy calibrations. Their dissemination through the adoption of a well-type chamber dosimetry system is furthermore described. Section 7 defines the dosimetry formalism employed for the determination of the dosimetry quantities used in this code of practice. In relation to Technical Reports Series No. 398, Absorbed Dose Determination in External Beam Radiotherapy [40] and Technical Reports Series No. 457, Dosimetry in Diagnostic Radiology: An International Code of Practice [41] that are based on the use of the beam quality correction factor k_{Q,Q_0}, in this code of practice a source model correction factor k_{sm,sm_0} is defined to take into account any difference between the actual source model sm and the one used for calibration, sm_0. The general procedure to properly perform brachytherapy dosimetry with the well-type chamber is given in Section 8, along with a description of methods to check for short and long term stability of the measuring system. Section 9 deals with the estimation of the uncertainties typically involved with the source strength measurement of LDR and HDR sources. The way measured reference quantities are useful in the clinical practice for assessing the dose to the patient is outlined in Section 10. The main brachytherapy source categories and treatment delivery methods are briefly approached.

Appendices are provided to complement the information given in the main body of the publication: Appendix I briefly mentions the antiquated quantities and units that are not recommended to be used any more for dosimetry purposes; Appendix II provides an insight into the present situation for dosimetry standards based on air kerma and absorbed dose to water for the sources considered in this code of practice; Appendix III provides a brief description of X-ray eBT devices and the current status of development of their dosimetry standards; Appendix IV provides an insight into some detector systems different from the well-type ionization chamber that might be used for brachytherapy dosimetry.

Appendix V describes in more detail the formalism found in the American Association of Physicists in Medicine (AAPM) Task Group No. 43 Report which is commonly used for dose distribution calculation in interstitial and intracavitary brachytherapy; Appendix VI introduces the theory for the estimation of measurement uncertainties.

2. BRACHYTHERAPY RADIOACTIVE SOURCES

Encapsulated radioactive sources for brachytherapy include many different designs and consist of radioactive materials permanently sealed in a capsule, or closely bonded and in a solid form [42]. The radioactive source is encapsulated in order to prevent escape or release of the radioactive material under normal conditions or in case of probable accidental events. Common materials used for brachytherapy source encapsulations are stainless steel, tungsten (W), titanium (Ti) and nickel (Ni). They provide adequate mechanical strength and low attenuation. In addition, the non-toxic material of the brachytherapy source housing is not interacting physically or chemically with body fluids, which could weaken the source integrity.

Photons are the most frequent type of radiation used in treatments, with energies ranging from 0.02 to 1.25 MeV. Low energy and high energy photons are distinguished as having an average energy less than or equal to 50 keV or exceeding 50 keV, respectively [23]. By design, the brachytherapy sources are positioned in proximity or within the target volume, temporarily or permanently. According to the definition provided by ICRU [43], LDR treatments show a dose rate to the dose prescription point (or surface) between 0.4 Gy h^{-1} and 2 Gy h^{-1}. High dose rate treatments are defined as those treatments delivering more than 12 Gy h^{-1} to the dose prescription point (or surface). Intermediate dose rates (2 Gy h^{-1} to 12 Gy h^{-1}) are in principle referred to as medium dose rates (MDRs); however, they are not commonly used in the clinical practice. Pulsed dose rate treatments mimic continuous LDR treatments by delivering small fractions of the prescribed dose — called pulses — with an MDR source. Regular pulses of 10 to 15 minutes duration are repeated once per hour until the prescribed fraction dose is reached.

2.1. MAIN PHOTON-EMITTING RADIOACTIVE SOURCES

There are currently six radionuclides used as photon emitting sources. With regard to their average energy, they can be grouped into low energy and high energy sources. Referring to the dose rate delivered to the dose prescription point (or surface), they can also be grouped into LDR, HDR and PDR sources. The division and type of photon emitting sources are shown in Table 1, with eBT sources being included for comparison purposes.

The low-energy photon-emitting sources using ^{103}Pd, ^{125}I and ^{131}Cs radionuclides have the advantage of being easily shielded as their average energies are approximately 20 keV, 28 keV and 30 keV respectively. Given the low energies and relatively short half-lives (see Section 3.4.1), these sources are mostly used for LDR permanent implants with sources ordered on a patient case-specific basis. In general, these low energy LDR sources are encapsulated in 0.8 mm diameter titanium tubes with lengths of approximately 5 mm; thus, the colloquial labelling of these sources as seeds.

The high-energy photon-emitting sources using ^{192}Ir, ^{137}Cs and ^{60}Co radionuclides have average energies of approximately 0.38 MeV, 0.66 MeV and 1.25 MeV, respectively. Their half-lives are given later in Section 3.4.1. These sources require the use of high Z shielding for close proximity work and are not permanently implanted due to their relatively long half-lives (see Section 3.4.1). For sources using ^{137}Cs, the radioactive material is contained within 3 mm diameter and 20 mm long stainless steel tubes for temporary LDR implants. The ^{192}Ir and ^{60}Co HDR sources are used for temporary applications and reused on multiple patients. The radioactivity is contained in a capsule having an outer diameter of approximately 1 mm and a length of 3–5 mm, which is attached to a source-drive wire for positioning by the HDR remote afterloader. PDR ^{192}Ir sources are similar in length and have a decreased activity compared to the standard HDR ^{192}Ir sources. They are also driven by a remote afterloader that is programmed to deliver the dose pulses with some inter pulse interval.

Dosimetric characteristics of sources are sensitive to the specific encapsulation geometry and internal radionuclide distribution. Particularly at low energies, self-absorption and filtration effects are significant, and contaminant photons due to the characteristic X-rays, which are produced in the outer layers of steel or titanium source encapsulations, need to be considered (see Section 3.1). Seed models using the same radionuclide and with relatively small differences in manufacturing processes and/or in their design may therefore show significant dosimetric differences. Even if less sensitive, high energy source models might also show different radiation emissions from one model to another. Approximate values for the average energy of the emitted photons can only be provided because the actual average energy is specific to each source model.

TABLE 1. DIVISION OF BRACHYTHERAPY PHOTON SOURCES

Energy	Dose rate	Type
LE[a]	LDR[c]	I-125, Pd-103 and Cs-131 seeds
HE[b]	LDR[c]	Cs-137 tube sources; Ir-192 wires, pins and needles
LE[a]	HDR[d]	eBT X-ray sources
HE[b]	HDR[d]	Ir-192 and Co-60 HDR afterloaders
HE[b]	PDR[e]	Ir-192 PDR afterloaders

[a] LE: Low energy (average energy $\leq$ 50 keV).
[b] HE: High energy (average energy > 50 keV).
[c] LDR: Low dose rate (dose rate > 0.4 Gy h^{-1} and < 2 Gy h^{-1}).
[d] HDR: High dose rate (dose rate < 12 Gy h^{-1}).
[e] PDR: Pulsed dose rate.

Globally, there are approximately two dozen low-energy and high-energy radionuclide-based photon-emitting sources currently on the market. LDR ^{137}Cs tubes are used rarely and with decreasing popularity, as are the alternative of PDR ^{192}Ir sources and ^{192}Ir LDR wires, pins and needles.

2.2. BETA-EMITTING RADIOACTIVE SOURCES

Beta emitting sources are used less frequently but have the potential for better dose conformity than photon emitters for shallow disease. These include ^{106}Ru (decaying to ^{106}Rh, which is also a beta emitter) and ^{90}Sr (decaying to ^{90}Y, which is also a beta emitter). Considering the influence of the daughter radionuclides on the beta energy and source half-life for ^{106}Ru and ^{90}Sr, the maximum energies are approximately 3.54 MeV and 2.27 MeV, respectively [27]. Their half-lives are given later in Section 3.4.1. Due to its high specific activity, ^{90}Sr/^{90}Y has been used as an HDR source for temporary intravascular brachytherapy (IVBT) treatments with thin sources administered for intracardiac BT via a catheter [44–46].

Beta emitters also have a long history in ophthalmology treatment, having the advantage of reduced dose penetration in tissue [47]. Sources of ^{106}Ru and ^{90}Sr used in intraocular brachytherapy are usually curved in the form of plaques and superficial applicators, respectively, conforming to the shape of the eye. Ophthalmic plaques containing ^{106}Ru/^{106}Rh are available in different diameters, and the active area is customized to different tumour sites [48]. They may have symmetrical and asymmetrical shapes with cut-outs for the optical nerve or iris. The dose is commonly prescribed at a set distance along the plaque central axis [28]. Existing ophthalmic ^{90}Sr/^{90}Y superficial applicators have a 10–18 mm diameter and are used for the surface treatment of anterior conjunctival lesions near the cornea (i.e. pterygium) [49–51].

These ^{90}Sr/^{90}Y applicators are, however, no longer being manufactured. If the user has a recently calibrated applicator (i.e. 3 to 5 years), it could be used clinically. The value would be given in absorbed dose rate to water (Gy s^{-1}) at a specified time, and the dose uniformity of the source would also be provided. The University of Wisconsin Accredited Dosimetry Calibration Laboratory (UWADCL) calibrated 222 sources from 1997 to 2008, and a large number of sources did not exhibit a suitable dose uniformity. Comparison to prior source strengths indicated an average discrepancy of −19% with values ranging from −49% to +42% [52]. If the user identifies an old source like this, the suggestion is to dispose of it, especially if it is without a recent calibration (i.e. within the last 5 years). In addition, it is recommended to leak test the source within one year according to international guidance.

There are significant obstacles that make the measurement of the source strength for ^{90}Sr/^{90}Y ophthalmic applicators and ^{106}Ru/^{106}Rh eye plaques challenging in the clinical setting. Recently, ^{106}Ru/^{106}Rh plaques have been calibrated with a concave windowless extrapolation chamber. A technique is provided in Hansen et al. [53] to calibrate an ionization chamber that can be used in a clinical situation. Prior to this, there was no recommended traceably-calibrated instrumentation to locally perform this type of measurement [28], and there is no international consensus on the calibration procedure, frequency and timing, nor on the instrumentation and procedures to be used in the hospital. This code of practice does therefore not apply to these sources. In these instances, alternative methods to measure the dose rate at a specific depth from the source are recommended. Examples are provided in the literature for a variety of detectors (i.e. extrapolation chambers, thermoluminescence dosimeters (TLDs), radiochromic films, plastic scintillators, silicon diodes, diamond detectors and small volume ionization chambers) [53–56]. Useful information can be found in Appendix IV.

2.3. OTHER PHOTON-EMITTING RADIOACTIVE SOURCES

Beyond the aforementioned sources, there are other BT sources that are innovative, but for which there may not be a robust means of providing source strength calibrations [57].

Ytterbium-169 has been proposed as a radionuclide to replace ^{192}Ir as an HDR source due to its lower mean photon energy of 0.09 MeV compared to 0.3 MeV [57, 58]. In practice, however, the amount of shielding reduction is not substantial, and its half-life (32 days) is shorter than that of ^{192}Ir (73.83 days), requiring more frequent source changes. Furthermore, there are no manufacturers offering this source and no primary calibration standards are available at present. There was a manufacturer in the past that offered a ^{169}Yb source with a traceable calibration [59].

Other innovative sources include seeds containing shielding to provide directional radiation [60–65], source-applicator combinations utilizing shielding for large-scale radiation directionality [66–69], and dynamic shielding of source-applicator assemblies to employ intensity modulation [70–75]. While many of these technologies are promising or already in clinical use, the clinical team needs to be cognizant of dosimetry and treatment planning challenges as outlined in the Task Group No. 167 Report of AAPM and European Society for Radiotherapy and Oncology (ESTRO) [57].

3. QUANTITIES AND UNITS

The relevant quantity for brachytherapy applications is the absorbed dose to water D_w measured at points that are clinically relevant and therefore close to the source. However, reference dosimetry of photon emitting sources has, for decades, been based on standards and transfer chambers in terms of air kerma, and availability of calibrations based on primary standards in terms of absorbed dose to water is currently limited [76–78].

Established dosimetry protocols for photon emitting sources are therefore based on reference air kerma rate (RAKR) or air kerma strength (AKS) standards and transfer instruments. Among these protocols, the AAPM TG-43 report and updates [79–83] play a prominent role as the underlying methodology that has been adopted worldwide. It is very important that the user is aware of the quantity used for the source strength. There have been mistakes made in putting the wrong quantity in the TPS [84].

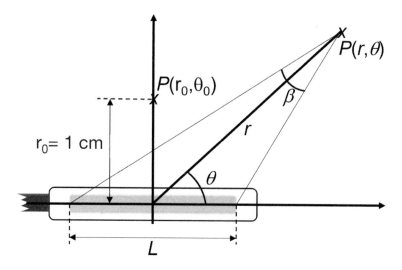

FIG. 1. The polar coordinate system chosen as the reference coordinate system for the AAPM TG-43 formalism. The radioactive content of the source is shown in grey (length L) and is surrounded by its encapsulation. θ_0 equals 90°.

The reference coordinate system chosen in this publication is the same as the one given in the AAPM TG-43 Report [79] and is shown in Fig. 1. According to this reference system, β is the angle (typically measured in radians) subtended by the point of interest $P(r, \theta)$ and the two ends of the source active element, θ is the polar angle between the source longitudinal axis and the ray from the centre of the source to the point of interest $P(r, \theta)$, and r is the distance (typically measured in cm) from the source centre to the point of interest $P(r, \theta)$. θ_0 is the reference polar angle and r_0 the reference distance from the source centre.

Table 2 provides a summary of the main physical quantities used in this code of practice, including their typical units.

3.1. REFERENCE AIR KERMA RATE AND AIR KERMA STRENGTH

According to the International Commission on Radiation Units and Measurements (ICRU) [43, 85–87], the RAKR $\dot{K}_{\delta,R}$ is defined as the air kerma rate due to photons of energy greater than a cut-off value δ, at a reference distance d_R of 1 m from the source centre, on the transverse plane normal to the long axis of the source and bisecting it, corrected for air attenuation and scattering as well as for possible photon scattering from any nearby walls, floors, and ceilings as well as from nearby objects in the room (i.e. in vacuo).

TABLE 2. PHYSICAL QUANTITIES USED IN THIS PUBLICATION, INCLUDING THEIR TYPICAL UNITS *(Physical quantities with no units are indicated with "–".)*

Quantity	Typical units	Short definition
$\dot{D}_{\mathrm{W}}(r,\theta)$	mGy·h^{-1} or Gy·h^{-1}	Absorbed dose rate to water at the point of interest P(r, θ)
$\dot{D}_{\mathrm{W,R}}$	mGy·h^{-1} or Gy·h^{-1}	Absorbed dose rate to water *in water* at the reference point P(r$_0$, θ$_0$); $\dot{D}_{\mathrm{W,R}} = \dot{D}_{\mathrm{W}}(r_0, \theta_0)$
δ	keV	Photon energy cut-off value used for air-kerma rate evaluation
$F(r, \theta)$	—	2D anisotropy function: ratio of the dose rate at distance r and angle θ, around the source, to dose rate on the transverse axis at the same distance r
$\varPhi_{\mathrm{an}}(r)$	—	1D anisotropy function: ratio of the dose rate at distance r, averaged over the entire solid angle (i.e. 4π), to dose rate along the transverse axis at the same distance r
$G_{\mathrm{X}}(r, \theta)$	cm^{-2}	Geometry function: considers both the inverse square law and the influence of the approximate physical distribution of the radionuclide on the dose distribution. The subscript X designates either P or L if the point source or line source approximation is chosen, respectively
$g_{\mathrm{X}}(r)$	—	Radial dose function: describes the dose rate at distance r from the source along the transverse plane relative to the dose rate at the reference distance r$_0$, excluding the dose geometrical fall-off effects modelled with the geometry function. The subscript X designates either P or L if the point source or line source approximation is chosen, respectively
$\dot{K}_{\delta}(d)$	mGy·h^{-1}	Air kerma rate at distance d from the source, in vacuo, due to photons of energy greater than δ

TABLE 2. PHYSICAL QUANTITIES USED IN THIS PUBLICATION, INCLUDING THEIR TYPICAL UNITS *(Physical quantities with no units are indicated with "−".)* (cont.)

Quantity	Typical units	Short definition
$\dot{K}_{\delta,R}$	$mGy \cdot h^{-1}$	Reference air kerma rate: air kerma rate, in vacuo, at a reference distance 1 m from the source centre, along the source transverse plane due to photons of energy greater than δ
k_{TP}	—	Air density correction factor
k_{leak}	—	Leakage currents correction factor (corrects for any measured signal not due to the source being measured)
k_{elec}	$A \cdot rdg^{-1}$ or $nC \cdot rdg^{-1}$	Electrometer calibration coefficient (i.e. rdg stands for readings on the electrometer display)
k_{pol}	—	Polarity correction factor
k_s	—	Ion recombination correction factor
k_{dec}	—	Source decay correction factor
k_{alti}	—	Altitude correction factor (in addition to the air density correction factor and specific for low energy sources)
k_{sm,sm_0}	—	Source model correction factor: corrects for the difference between the response of an ionization chamber irradiated with the calibration source model sm_0 and with the user source model sm. Both sm_0 and sm contain the same type of radionuclide but are assumed to have different geometries. The source model correction factor also depends on the type of well-type chamber and source holder
Λ	cm^{-2}	Dose rate constant: absorbed dose rate at the reference point $P(r_0, \theta_0)$ per unit of S_K, after having removed geometry function effects
Λ_{r_0}	—	Notation indicating that the dose rate constant is defined as the absorbed dose rate at the reference point $P(r_0, \theta_0)$ per unit of $\dot{K}_{\delta,R}$, after having removed geometry function effects

TABLE 2. PHYSICAL QUANTITIES USED IN THIS PUBLICATION, INCLUDING THEIR TYPICAL UNITS *(Physical quantities with no units are indicated with "−".)* (cont.)

Quantity	Typical units	Short definition
M_{sm}	A	Reading of the dosimeter irradiated with a source model sm, corrected for the influence quantities other than the source model (e.g. air density, leakage currents, electrometer, polarity, and ion recombination)
$M_{sm,\,raw}$	A or rdg	Raw reading of the dosimeter irradiated with a source sm, uncorrected for any influence quantity
$N_{\dot{K}_{\delta,R},\,sm_0}$	cGy·h^{-1}·A^{-1}	Reference air kerma rate calibration coefficient measured with the calibration source model sm_0
$N_{S_K,\,sm_0}$	cGy·cm^2·h^{-1}·A^{-1}	Air kerma strength calibration coefficient measured with the calibration source model sm_0
P	kPa	Ambient pressure
RH	%	Ambient relative humidity
S_K	1 cGy·cm^2·h^{-1} = 1 U	Air kerma strength: air kerma rate $\dot{K}_\delta(d)$, in vacuo, due to photons of energy greater than δ at any distance d from the source centre, on a transverse plane normal to the long axis of the source and bisecting it, multiplied by the square of the distance d^2
$t_{1/2}$	d or y	Half-life of the radionuclide
T	°C	Ambient temperature

The SI unit for $\dot{K}_{\delta,R}$ is Gy·s^{-1} at 1 m from the source, but for the purposes of source specification it is usually more convenient to use mGy·h^{-1} and Gy·h^{-1} for LDR and HDR brachytherapy sources, respectively. The formalism behind the definition of the RAKR is equal to that of the definition of the absorbed dose to water in EBRT [40], and the quantity itself is defined under reference conditions.

According to the definition given by the AAPM [37, 79, 81], the air kerma strength (AKS) S_K is defined as the air kerma rate $\dot{K}_\delta(d)$ (in vacuo, due to photons of energy greater than δ) at any distance d from the source centre, on a transverse plane normal to the long axis of the source and bisecting it, multiplied by the square of the distance d^2:

$$S_K = \dot{K}(d)d^2 \qquad (1)$$

RAKR or AKS measurements are performed at any distance much larger than the linear length L of the radioactivity distribution of the source core. This allows the application of the point source approximation and can be assumed for d being at least ten times bigger than L. The unit for S_K is $\mu Gy \cdot m^2 \cdot h^{-1}$ which is denoted by the symbol U where $1\ U = 1\ \mu Gy \cdot m^2 \cdot h^{-1} = 1\ cGy \cdot cm^2 \cdot h^{-1}$. The units for $\dot{K}_{\delta,R}$ and S_K are therefore different.

For both $\dot{K}_{\delta,R}$ and S_K, the energy cut-off δ is defined to exclude low energy or contaminant photons, since they increase the air kerma rate without contributing significantly to the absorbed dose at depths that are clinically relevant (i.e. >1 mm in tissue). Low energy photons typically originate in the outer layers of the metallic source capsule as the result of the interaction of the emitted source core radiation with the shielding (e.g. characteristic X-rays). Values for δ depend on the intended clinical application and are typically 5 keV and 10 keV for low and high energy photon-emitting sources, respectively.

The physical difference between these two quantities is that $\dot{K}_{\delta,R}$ is always defined at 1 m, whereas S_K incorporates the distance and the inverse square law as applied to an isotropic point source. This difference is important, since in some cases the distance of measurement d_m is different from the reference distance of $d_R = 1$ m. In that case, a correction factor is applied given by the inverse square law at the measurement distance according to:

$$\dot{K}_{\delta,R} = \dot{K}_\delta(d_m)\left(\frac{d_m}{d_R}\right)^2 \qquad (2)$$

3.2. ABSORBED DOSE TO WATER AND THE DOSE RATE CONSTANT

According to the formalism provided in the AAPM TG-43 report (and updates) [79–83] conversion from S_K to the absorbed dose rate to water at the reference point P(r_0, θ_0), $\dot{D}_{W,R} = \dot{D}_W(r_0,\theta_0)$ is obtained by multiplying S_K with the dose rate constant Λ, which is defined as the absorbed dose rate to water at the reference point per unit of S_K according to:

$$\Lambda = \frac{D_W(r_0, \theta_0)}{S_K} \tag{3}$$

Analogously, conversion from $\dot{K}_{\delta,R}$ to $\dot{D}_{W,R}$ in the same reference conditions can be obtained using the dose rate constant Λ_{r_0} [88], which is defined as:

$$\Lambda_{r_0} = \frac{D_W(r_0, \theta_0)}{\dot{K}_{\delta,R}} \tag{4}$$

Λ and Λ_{r_0} are characteristic of the radionuclide and of the particular source model. For photon sources the reference point is usually specified at a distance $r_0 = 1$ cm along the transverse plane of the source ($\theta_0 = 90°$) (see Fig. 1 for the reference coordinate system). Consensus data for the dose rate constants of the main commercially available sources at the time of this report are available in the literature [23, 79–83]. The unit for Λ is usually cm^{-2} whereas Λ_{r_0} is dimensionless. Current source data meeting the AAPM prerequisites are also available on-line on the joint AAPM/IROC Houston Brachytherapy Source Registry [89]. Source data may also be found on other on-line resources [90, 91].

3.3. RECOMMENDED CALIBRATION QUANTITIES

This section describes the recommended quantities for the specification of the strength of brachytherapy sources this code of practice deals with. All of these source types may be calibrated using a re-entrant well-type ionization chamber with a specific source-positioning holder to position the source at the centre of the chamber.

3.3.1. Photon-emitting radioactive sources

The RAKR $\dot{K}_{\delta,R}$ is recommended by ICRU [43, 85–87] as the reference quantity for the source strength specification of photon-emitting radioactive sources. The AKS S_K is recommended by the AAPM [37, 79, 81] and National Institute of Standards and Technology (NIST). The quantity S_K is currently inserted into most of the available TPSs to perform dose distribution calculations.

3.3.2. Beta-emitting radioactive sources

Unlike photon-emitting radioactive sources, source strength of beta emitters is not specified in terms of S_K or $\dot{K}_{\delta,R}$. The recommended quantity for the strength specification of beta emitting sources is the absorbed dose rate to water at a reference distance r_0 in water from the external surface of the source $\dot{D}_W(r_0)$, along the axis of symmetry of the source. The recommended reference distance of calibration for IVBT sources is $r_0 = 2$ mm [28]. Even if measurements at this short distance are challenging, this distance is chosen considering the shallow penetration, the relevance to clinical applications "and the difficulty of accurate dose determination on the surface of the sources" [87].

3.4. NUCLEAR DECAY: HALF-LIVES AND DATE AND TIME STANDARD

An incorrect half-life $t_{1/2}$ or its improper update have been among the main sources of errors and treatment misadministration in brachytherapy in the past [11, 12, 92]. In fact, it is recommended to calculate the source activity accurately and to keep updated according to the date and time of a performed measurement or treatment, since the dose rate delivered to the patient is proportional to it.

3.4.1. Reference half-lives

Half-life is specific for each radionuclide and is fundamental to determine the activity of the source at the time of treatment. Reference $t_{1/2}$ data can be found in the literature [93–97] and in resources available on-line [98–102]. Recommended half-lives of some of the radionuclides used in brachytherapy are provided in Table 3. For unit conversion from years (y) to days (d), the factor 365.242198 d y^{-1} is applied, and rounded where appropriate [103].

3.4.2. Reference date and time standard

Since date and time are fundamental to provide an accurate evaluation of the current source activity, it is important to use a proper standard to define them. To this end, the International Organization for Standardization (ISO) issued the 8601:1-2019 standard to provide a consistent convention for the representation of numeric dates and times and their exchange between countries [104]. The recommended representation for calendar dates and times of day is given in Table 4. The recommended time standard is the Coordinated Universal Time

(UTC), which is the time standard commonly used across the world and from which local time is derived.

TABLE 3. RECOMMENDED HALF-LIVES FOR RADIONUCLIDES USED IN BRACHYTHERAPY

Radionuclide	Element name	Atomic number Z	Main decay used for brachytherapy	Half-life $t_{1/2}$
Co-60	Cobalt	27	γ	1925.21 ± 0.29 d [95]
Sr-90	Strontium	38	β	28.80 ± 0.07 y [95]
Rh-106	Ruthenium	44	β	371.5 ± 2.5 d [97]
Pd-103	Palladium	46	γ	16.991 ± 0.019 d [99, 100]
I-125	Iodine	53	γ	59.388 ± 0.028 d [96]
Cs-131	Caesium	55	γ	9.689 ± 0.016 d [99, 100]
Cs-137	Caesium	55	γ	30.05 ± 0.08 y [95]
Ir-192	Iridium	77	γ	73.827 ± 0.013 d [94]

TABLE 4. RECOMMENDED REPRESENTATION FOR LOCAL CALENDAR DATE AND TIME (according to [104])

	Calendar date	Time of day
Extended format	YYYY-MM-DD	h:m:s
Specific format (e.g. 24 November 2010, 5 minutes and 30 seconds past 21 hours)	2010-11-24	21:05:30

Note: Digit used to represent characters in the time scale component: Y: calendar year; M: calendar month; D: calendar day; h: clock hour: m: clock minute; s: clock second.

4. INSTRUMENTATION

4.1. THE RE-ENTRANT WELL-TYPE IONIZATION CHAMBER DOSIMETRY SYSTEM

As it is thoroughly discussed later in this publication, the recommended method to measure the strength of the main brachytherapy sources is based on the use of a re-entrant well-type ionization chamber, usually called a *well-type chamber*. The system for brachytherapy dosimetry is considered as the combination of the following components:

(a) A vented well-type chamber;
(b) A source holder to position the source inside the well-type chamber;
(c) An electrometer;
(d) An extension cable (if required).

A description of each one of these components can be found below. Since an important part of the dosimetry system is also a method for well-type chamber constancy checking, recommended additional accessories and methods are discussed in Section 8.5.

The system needs to be stored in a suitable location (particularly with regard to security and environmental control) and should be used by authorized personnel only. It is also recommended in this code of practice to bring the measurement assembly to the site before starting the measurement since it requires time for the dosimeter to equilibrate with the environment. Since well-type chambers are heavier and bigger than other smaller measurement devices such as thimble-type ionization chambers, longer times are generally needed to equilibrate. The system necessitates a warm-up period before commencing any measurements. Users need to refer to the relevant manual for recommended warm-up times.

4.1.1. The well-type chamber

The recommended well-type chamber is of the type designed for brachytherapy dosimetry applications and able to be used to derive the source strength of LDR and HDR brachytherapy sources. It is recommended that only vented (open to atmosphere) type chambers are used. Sealed chambers are not recommended for measurements as over time they may start leaking, which may cause a change in their calibration coefficients. Specifically, pressurized well-type chambers commonly used in nuclear medicine applications are not to be used for brachytherapy measurements because they have calibration settings for nuclear medicine radionuclides and not for brachytherapy sources and provide readings in units of activity. The use of a chamber from a nuclear medicine practice may also lead to contamination of the source holder and/or chamber.

A cross-sectional schematic of a typical well-type chamber is depicted in Fig. 2. There is a cylindrical outer chamber wall with an inner wall that delimitates an opening for inserting and positioning the source inside the well at a distance from the bottom of the chamber. A removable source holder, also included in the sketch, is used to achieve this. The well-type chamber has three electrodes, like other ionization chambers. The triaxial cable is connected to an electrometer which measures the ionization current and supplies a high voltage between the central collecting electrode and the outer electrode. The guard electrode is sandwiched between the outer and collecting electrodes and separated from both by a high voltage insulator. The electric potential of the guard electrode is always the same as that supplied to the collecting electrode, to ensure that the collected charge does not leak through the insulator to the environment. The voltage gradient between the outer and collecting electrodes defines the sign of charge (negative or positive) collected with the electrometer.

An ideal chamber would show no sensitivity to the position of the brachytherapy source within the well, but all practical chambers generally have what is referred to as a 'sweet spot' where the chamber signal is the maximum. The length of the sweet spot, defined for a single source as the full width at 95% of the maximum signal, has to be as large as possible, and certainly larger than the longest dimension of the source being measured.

4.1.2. Well-type chamber source holders

A source holder is used to establish a reproducible source position within the chamber cavity. Since a source holder is part of the calibration chain influence quantities, the measurements need always to be carried out with the appropriate holder made by the manufacturer for that particular source model and well-type chamber. Holders of the same model are not to be swapped between different well-type chambers, otherwise the well-type chamber calibration will be compromised. To ensure repeatable measurements, some source holders have a marking on the upper surface that needs to be rotationally aligned with labelling on the body of the well-type chamber.

HDR and LDR source holders are different, and they are not to be mixed. The most common material used for HDR source holders is either polymethyl methacrylate (PMMA) or a low Z metal. In some cases, a low mass thermal insulator (usually Styrofoam) is added [105]. Encircling the central aluminium tube, the insulator maintains thermal equilibrium between the ion collecting

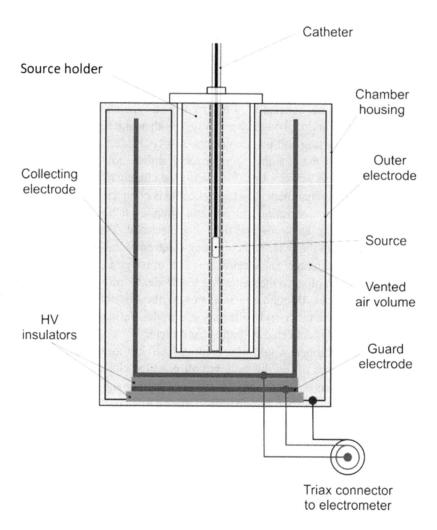

Catheter

Source holder

Chamber
housing

Collecting
electrode

Outer
electrode

Source

Vented
air volume

HV
insulators

Guard
electrode

Triax connector
to electrometer

FIG. 2. Cross section of a generic design of a well-type chamber. Actual well-type chambers
from different manufacturers might look different. A different source holder without a catheter
is used with LDR seeds.

volume and the source during the measurement, since heat can be generated by
the higher activity of HDR sources. In the case of LDR seeds, the source holder
is composed of a central plastic tube fastened to a low mass, low Z frame to
consistently position the seed(s) within the centre of the well-type chamber.

4.1.3. Electrometers, cables and connectors

An electrometer is used for measurements of ionization current and charge. Different modes, like continuous and triggered charge collection, are usually available on the electrometer. Some modern electrometers are current-sensing devices, rather than charge-sensing, and therefore it may be preferable to measure current. It is therefore important to understand the preferred measurement mode of the electrometer being used.

It may be possible to set the polarity of the polarizing voltage that is provided by the electrometer, so that the well-type chamber can be operated with the same voltage gradient between the inner (collecting) electrode and the outer electrode that was used during calibration at the calibration laboratory. If the user of the well-type chamber measures the same sign of charge/current (either negative or positive) that was measured at the calibration laboratory, there is no need to apply a polarity correction factor.

It is also important that the variation of the voltage provided by the electrometer is possible, in order to determine the ion-recombination correction factor, k_s (i.e. the reciprocal of the ion collection efficiency) [36, 106].

Other important parameters relevant for the choice of electrometers are the current/charge measurement range and resolution, linearity and zero drift. It is recommended that the electrometer connected to the ionization chamber is suitable for measurements of ionization currents up to 200 nA for HDR sources and has a resolution at the femtoampere level for LDR sources. The majority of commercial electrometers currently available meet these requirements but testing and investigation may be required for older instruments to verify compliance.

4.1.4. Connectors and extension cable

An important part of the measurement assembly is constituted by the electrometer and well-type chamber connectors. Most commonly, dust collects on the interior part of a connector. The connector cleaning procedure should be periodically conducted, or it can be done whenever there are signs of drift or leakage. After the visual inspection of a connector, dry, oil-free compressed air can be used to remove dust and contaminants.

In many cases, an extension cable is needed to connect the well-type chamber and electrometer, and additional visual inspection and leakage checks are carried out to ensure the correct operation of the cable.

4.2. REFERENCE-CLASS WELL-TYPE IONIZATION CHAMBERS

Detectors used for the calibration of brachytherapy sources need to meet a minimum level of performance so that operation of the instrument does not negatively impact the measurement procedure. Specifications have been developed for reference-class ionization chambers used for EBRT [107–109] and therefore it is appropriate to develop a similar specification for well-type chambers.

4.2.1. Specification of reference-class well-type chambers

(a) Open to atmosphere: Only unsealed, unpressurized chambers are recommended for brachytherapy measurements.

(b) Size of the well: The chamber needs to be large enough to accommodate the source to be measured while approximating a 4π geometry to minimize the impact of source rotational orientation within the well. For a single source, this implies a well depth of at least 100 mm and a well diameter of around 30 mm.

(c) Leakage current: Leakage is defined as the signal measured in the absence of a source within the well of the ionization chamber. This is more likely to be electrical in nature (i.e. leakage currents arise between conductors within the ionization chamber, connecting cable or electrometer), but may be also due to external radiation sources (due to the high sensitivity of the chamber). In either case, the leakage signal is recommended to be less than 0.1% of the reading obtained with the source present (for an HDR/PDR source). Without any source in place, the leakage current is advised to be <50 fA without great positive and negative variation.

(d) Sweet spot and sweet spot length (axial positional response): Given standard source holders and the cylindrical symmetry of the chamber design, the only variable associated with the sweet spot is the longitudinal position from the bottom of the well. Although, intuitively, one would conclude that the position of the sweet spot is constant, there can be some source-to-source variability. The length of the sweet spot, defined for a single source as the full width at 95% of the maximum signal, has to be as large as possible, with a suggested length larger than 50 mm. A minimum sweet spot length of 30 mm, or larger than the longest dimension of the source being measured is recommended. For seed trains, a longer sweet spot length of at least 100 mm may be required.

(e) Ion recombination and polarity corrections: In general, ion-recombination and polarity corrections are not significant when measuring photon-emitting brachytherapy sources. Ion recombination is advised to be less than 0.2%. A polarity correction is not required if the user operates the well-type chamber with the same voltage setting and polarity that was used at the calibration laboratory.

(f) Signal magnitude (sensitive volume) and dynamic range: These two components are linked in that they are both a function of the sensitive detection volume of the chamber and the specification of the electrometer. For a reference class instrument, one desires a single system that can measure both HDR and LDR sources with similar accuracy. This is generally achieved with an ionization chamber collecting volume of approximately 200 cm^3 (at normal pressure, 101.325 kPa) combined with an electrometer able to measure currents in the range 200 pA to 200 nA with similar accuracy and precision. In addition, electrometers have digital displays. A minimum resolution of 0.1% of the typical reading is essential to avoid digitization errors impacting the measurement.

(g) Energy response: An ideal detector would show no sensitivity to the range of energy of photons emitted by the radioactive sources compared to the calibration energy. An open-to-atmosphere design allows the thin walls necessary to minimize attenuation of low energy photons. But given the variation in response as a function of energy, it is not possible to eliminate energy dependence. For open-to-atmosphere chambers, with a thin inner wall, the energy dependence will not have any significant impact on the measurement, since the use of source model-specific calibration coefficients is recommended. For thicker walled chambers further characterization by the user is required to ensure that any small source-to-source variations are not amplified by the large wall attenuation.

(h) Environmental sensitivity: The impact of environmental parameters (i.e. temperature T, pressure P, relative humidity RH) on air density changes are taken into account using standard methods for ionization chambers. The ambient humidity can cause significant effects (variation in leakage currents, mechanical stability (swelling), etc.). The impact of relative humidity on well-type chambers has been investigated [110] and the response of a reference-class chamber is supposed to not vary by more than 0.3% for the range 15% < RH < 80%.

(i) Short term repeatability: Following the recommended warm up period, an ideal chamber would show no variation in response from sequential insertions of a given brachytherapy source. Any variation in signal as a function of time, other than that caused by the decay of the source, will impact the measurement. This has been investigated extensively for cylindrical and parallel-plate ionization chambers used in EBRT but there are limited data for well-type chambers. However, it is reasonable to use a similar specification as in [107, 111], that the signal from the ionization chamber is supposed to stabilize within 10 minutes, and that the difference between initial and equilibrium readings will not be greater than 0.5%. After stabilization, the short term repeatability is advised to be within 0.1%. Repeatability can mean different things, but in this case, it refers to the standard deviation of a set of repeated readings for both a source fixed in position and when it is moved back and forth. For LDR BT dosimetry, the standard deviation related to short term repeatability can be higher.

(j) Long term stability: Well-type chambers, if well maintained, have demonstrated very high stability of their responses over many years, showing a standard deviation in repeat calibration coefficients less than 0.15% for HDR ^{192}Ir sources [112, 113]. The stability in the electrometer response is advised to be within 0.2% over two years [59, 114].

(k) Source holder: The source holder for each source model is a fundamental part of the brachytherapy dosimetry system. Only with the source holder adequate for the measured source, can the well-type chamber be considered as reference class and operating correctly. The combination of the well-type ionization chamber and the source holder is calibrated for a given source model. For well-type chambers with universal source holders where a flexible plastic catheter or steel needle needs to be pushed into the central borehole of the source holder, the catheter or needle become part of the secondary standard system. The calibration certificate should contain a description of the measurement set-up that was used at the calibration laboratory. It is essential that the well-type chamber is used with the equivalent accessories. Using different types of source holders and/or catheters or needles might invalidate the calibration coefficient stated on the calibration certificate.

4.2.2. Available reference-class well-type chambers

Based on the specification given in the previous section, it has been determined that the currently available chambers that can be regarded as reference-class instruments are those provided in Table 5. These chambers are suitable for use in both calibration laboratories and hospitals.

Well-type chambers having longer active lengths, such as the Standard Imaging IVB 1000, were specifically designed for measurements of the source strength of long source trains typically found for intravascular sources. Their properties in terms of volume, sensitivity, axial positional response and location of the chamber axial point of maximum response are different from those of well-type chambers designed for single sources.

Chambers that were previously manufactured and might be still in operation, which are considered to meet the specification, include the PTW 33004, also distributed with the name Nucletron SDS Type 077.09X. Its technical specifications are also provided in Table 5. Other chambers may be considered reference class if their performances can be shown to meet the above specification.

Manufacturers of well-type chambers are advised to provide the following information about their products to the user:

(a) The dimension of the active volume and whether it is vented to the atmosphere;
(b) The measuring range in RAKR (or S_K) for each possible radionuclide that might be measured;
(c) The polarizing voltage range;
(d) The working environment requirements (temperature, pressure, relative humidity);
(e) The main electrical characteristics (e.g. sensitivity, leakage, stability, wiring and cable connection);
(f) The source models and main characteristics of the source holders, and the approximate distance of the sweet spot from the base of the chamber well.

4.2.3. Commissioning of well-type chambers

For the models listed in Table 5, it is not necessary to establish the performance of a particular well-type chamber against the specification listed in Section 4.2.1. However, it is still necessary to carry out commissioning measurements to be confident that the detector is operating as expected and is 'fit for purpose'. It is important to note that not all the tests are needed before the chamber/electrometer/holder system can be used. For example, it is not realistic to establish long term repeatability prior to putting the system into service; however, such monitoring needs to be part of the ongoing quality control (QC) procedures of the SSDL/hospital (Section 8.5). Moreover, well-type chambers will be independently commissioned for all the different source types that are used in the calibration and clinical routine.

In Table 6 the recommended commissioning tests for the well-type chamber and the related tolerance levels are provided [116]. Results of the evaluated tests are supposed to be consistent with manufacturer specifications, particularly if these are more restrictive than the reported tolerance levels. Further investigation is required if the given limits are exceeded.

As an additional commissioning test, to provide further confidence that the system is working properly, it is recommended to measure the source strength applying the known calibration coefficient of the commissioned well-type chamber. The measured source strength needs to be compared to the one given in the source certificate, corrected for the radioactive decay. The difference is

TABLE 5. TECHNICAL SPECIFICATIONS OF REFERENCE-CLASS WELL-TYPE CHAMBERS GIVEN BY THE MANUFACTURERS

	Standard Imaging IVB 1000	Standard Imaging HDR 1000 Plus	PTW 33005 Sourcecheck$^{4\pi}$	PTW 33004[a,b,c]
Active volume (cm^3)[d]	475	245	116	164
Typical polarizing voltage (V)	300	300	400	400
Nominal sensitivity[e]	2.1 pA U^{-1} (Cs-137) 2.2 pA U^{-1} (HDR Ir-192) 2.4 pA U^{-1} (LDR Ir-192) 4.3 pA U^{-1} (I-125) 2.3 pA U^{-1} (Pd-103)	2.0 pA U^{-1} (Cs-137) 2.1 pA U^{-1} (HDR Ir-192) 2.3 pA U^{-1} (LDR Ir-192) 4.3 pA U^{-1} (I-125) 2.1 pA U^{-1} (Pd-103)	125 fA MBq^{-1} (Ir-192) 65 fA MBq^{-1} (I-125)	120 fA MBq^{-1} (Ir-192)
Typical axial response[f]	±0.3% over 100 mm	±0.5% over 25 mm	<3% over ±17.5 mm (Ir-192) <3% over ±20 mm (I-125)	<3% over ±17.5 mm (Ir-192) <3% over ±19.5 mm (Co-60)
Stability	0.2% over 2 years	0.2% over 2 years	≲±1 % per year	≲±1 % per year
Leakage current (fA)	<50	<50	≤50	<500

[a] Also distributed by Nucletron/Elekta as Source Dosimetry System (SDS) 077.09X.
[b] Not commercially available.
[c] Some unexpected behaviours, including a significant undesired detection volume, have been discovered for this well-type chamber. The detector can be used as a transfer instrument for source calibration; however, appropriate precautions are advised to be taken for optimal use [115].
[d] Vented to the atmosphere.
[e] 1 U = 1 μGy·m^2·h^{-1}.
[f] Around the sweet spot of the well-type chamber and along its central axis.

typically advised to be <3% for HDR/PDR sources and <5% for LDR seeds[1]. Larger discrepancies are advised to be investigated. The measured source strength and the source strength specified in the source certificate is advised to agree within their expanded uncertainties ($k = 2$).

Details about the electrometer commissioning are reported elsewhere [117]. For chambers not listed in Table 5, a wider investigation is required to determine reference-class performance and therefore suitability for determination of RAKR.

4.3. HDR BRACHYTHERAPY DELIVERY EQUIPMENT

An HDR brachytherapy afterloader is a computer-controlled system that can drive a high activity source (e.g. 370 GBq of ^{192}Ir or 74 GBq of ^{60}Co) from a shielded safe to a specific point in an applicator and then retract the source back into its safe after a predetermined dwell time. Afterloaders are operated from a computer-controlled console and some manufacturers offer additional control panels with touch screens, for easy and quick control command execution (see Fig. 3). An additional important function of the control console is to keep the source strength value updated and consistent with the actual source strength of the installed source.

In afterloaders that use stepping source technology, a single source, typically laser welded at one end to a source cable, moves in pre programmed steps through the applicators. An indexer of the afterloader directs the source cable from the safe to one of the openings/channels on the front surface of the unit. Several transfer tubes (e.g. up to 40) can be connected to the afterloader. The computer drives the source from the safe in the afterloader through a given channel to the programmed position in the applicator. These positions are known as dwell positions and a source can dwell there for a predefined amount of time, called the dwell time. The dwell positions and times in each channel are fully programmable, thereby giving a high level of flexibility of dose delivery in clinical brachytherapy applications. It is recommended to update the source dwell times to take into account the source decay.

The source safe material is usually tungsten. Typical maximum storage capacity for HDR ^{192}Ir and ^{60}Co sources needs approval by authorities for marketing and is determined by the regulatory requirements in different countries. When a source is in the safe, the dose rates at 1 m distance from the head of the unit are <1 μSv/h and <10 μSv/h for ^{192}Ir and ^{60}Co sources, respectively.

Prior to each source cable extension, the stepper motor will drive a check cable (with a dummy source at its end) into the programmed channel to verify the

---

[1] It is advised to the average of the measured RAKR of at least 5 LDR seeds.

TABLE 6. COMMISSIONING TESTS OF THE WELL-TYPE CHAMBER AND ASSOCIATED TOLERANCE LEVELS FOR REFERENCE-CLASS (RESULTS FOR THE EVALUATED TESTS ARE SUPPOSED TO BE CONSISTENT ALSO WITH MANUFACTURER SPECIFICATION)

Test	Tolerance
Mechanical integrity Can be verified by a physical check; radiographic images of the system might also help to possibly detect internal damage or loose cable connections.	No damage or loose cable connections; free vent hole.
Leakage current Should be measured with the method described in Section 8.2.3.4. The extension cable will also be verified if it is used in the standard measurement conditions. Due to the low ionization current from LDR sources, it is recommended that no extension cable be used for such sources.	<0.1% of the signal for an HDR/PDR source; <1% for an LDR source.
Sweet spot length For its quantification, the procedure for the sweet spot determination will be followed (see Section 8.2.1).	>3 cm (or larger than the length of the source being measured) for HDR/PDR/LDR brachytherapy sources.
Ion recombination Should be measured with the source dwelling in the sweet spot using the method described in Section 8.2.3.7.	≤0.2% for HDR/PDR/LDR brachytherapy sources.
Polarity effect Should be measured with the source dwelling in the sweet spot using the method described in Section 8.2.3.6.	If the chamber is operated at the same polarizing voltage and voltage gradient as used during calibration at the calibration laboratory, no polarity correction needs to be applied by the user. This is only required if the polarizing voltage and voltage gradient have the opposite sign compared to the settings used at the calibration laboratory.

TABLE 6. COMMISSIONING TESTS OF THE WELL-TYPE CHAMBER
AND ASSOCIATED TOLERANCE LEVELS FOR REFERENCE-CLASS
(RESULTS FOR THE EVALUATED TESTS ARE SUPPOSED TO BE
CONSISTENT ALSO WITH MANUFACTURER SPECIFICATION) (cont.)

Test	Tolerance
Short term repeatability Can be quantified as the standard deviation of a minimum of ten charge measurements, with the source dwelling in the same source position for the same dwell time (see Section 8.4). Two different tests need to be performed. (a) Measurements are repeated, and the source is not removed between one measurement and the other. (b) Measurements are repeated, and the source is removed and put/transferred back between one measurement and the other.	<0.1% for an HDR/PDR source[a]; <1% for an LDR source[a].

[a]   The tolerance values both apply for tests (a) and (b).

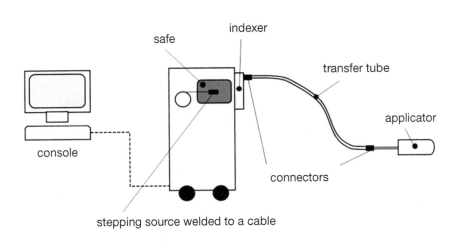

FIG. 3.   Simplified drawing of HDR brachytherapy delivery equipment connected to a (cylindrical) applicator.

integrity of the system. This system can detect a transfer tube and/or applicator obstruction or constriction from increased friction in the cable movement. For closed end applicators, the check source will reach the first possible dwell position and extend a bit further to detect the applicator end. This additional verification is performed to check possible erroneous settings for the most distal source dwell position.

Under certain fault conditions, such as the stepper motor failing to retract the source, a high torque, direct-current emergency motor will retract the source.

HDR afterloaders from various manufacturers differ in the following features:

(a) The method of source movement is either from the most distal position in the applicator backwards or starting at the most proximal position first and then distally towards the last programmed dwell position. The source speed can be up to 0.5 to 0.6 m s^{-1}.

(b) The possible range of positions over which the source can dwell in an applicator/catheter (e.g. is at least 400 mm).

(c) The number of definable dwell positions, source step size and dwell times (e.g. definable in 0.1 s increments in the range 0.1 to 999.9 s per dwell position).

(d) The number of guaranteed source extensions and retractions. Typically, about 10 times more applies to ^{60}Co sources if compared to ^{192}Ir.

The safety of afterloaders is ensured through the proper functioning of the following components:

(a) Independent backup retraction system/motor in the event of primary retraction system failure;

(b) Additional manual hand crank to retract the source in an emergency;

(c) Measurement system to detect the retraction of the source to the safe with an in-unit radiation detector (connected to a visual and audible indicator) and at least either source motion detectors, or switches/photoelectric barriers;

(d) Additional independent room survey monitoring system;

(e) Secondary timer, encoders for motion detection and source position verification;

(f) Functional emergency buttons on the console and at the walls both inside and outside the room;

(g) Door interlocks;

(h) Access to the control console protected by password and hardware key.

A handheld radiation survey meter should also be available as an important additional part of the safety system as it allows to independently check the source retraction after the irradiation.

Transfer tubes are manufacturer-specific plastic tubes that, when attached to an indexer or channel of an afterloader, are used to transfer the source from the safe into the applicators for irradiation. For source strength measurements, they are fundamental to connect the afterloader to the source holder placed inside the well-type chamber. Transfer tubes and source holders are therefore an important, integral part in the brachytherapy source strength measurement chain. In a clinical environment, in addition to the source holder, a dedicated transfer tube should be kept exclusively for source dosimetry.

On both ends of a transfer tube there is a connector to allow for proper safe attachment. Some transfer tubes have a ball bearing that blocks the path of the source if no applicator is attached. In some cases, connection is encoded to avoid possible misconnections. The accuracy of source positioning depends on the tube integrity. Some transfer tubes allow for small adjustments of their length.

During measurements, the tubes are guided from the afterloader indexer to the applicator as straight as possible since bending of the transfer tubes may lead to their damage. After detaching them from the afterloader, bending and damage can be prevented by storing the tubes elongated in a horizontal position or hanging them vertically in a specially designed tube holder.

Acceptance testing of the HDR brachytherapy delivery equipment verifies that the treatment unit meets safety standards and any specific regulatory requirements in the treatment and control room (e.g. mechanical and electrical safety features of the system, safety interlocks, shielding properties, emergency functionalities, radiation surveys of the afterloader). Moreover, the acceptance procedure verifies that measured parameters satisfy the manufacturer's contractual specification (e.g. positional accuracy, timing accuracy, integrity and activity of the source).

After acceptance, comprehensive commissioning of new HDR brachytherapy delivery equipment has to be implemented. It is advised to repeat commissioning in case significant hardware or software updates of the brachytherapy system are performed. In both SSDLs and hospitals, important steps in commissioning comprise (but are not limited to):

(a)   Development of operational and QA/QC procedures;
(b)   QA of the HDR brachytherapy delivery equipment, including verification of the proper application of the source decay correction at the treatment console;
(c)   Training of the involved staff.

Information on how to implement a comprehensive QA programme can be found elsewhere [10, 37, 118–123].

4.4. INSTRUMENTS FOR AIR DENSITY AND RELATIVE HUMIDITY MEASUREMENTS

For all air ionization chambers vented to the atmosphere, it is necessary to consider variations in the density and/or humidity of the air inside the sensitive volume. Since it is generally not possible to carry out a direct measurement, it is deemed sufficiently accurate to measure the analogous parameters, specifically the temperature of the well-type chamber housing and the humidity of the surrounding air. Putting the temperature probe either inside the well without the source holder being present, or in close proximity to the outside of the well-type chamber housing, will result in a good estimate of the air temperature inside the sensitive volume. A measure of the room air temperature is sufficient only if the well-type chamber is in complete thermal equilibrium with the room, something that is challenging to verify and therefore not a recommended approach. The same recommendations are given for both SSDLs and clinical situations and this is for two reasons:

(a)   The widespread availability of accurate and affordable temperature, pressure, and humidity sensors;
(b)   Uniformity of equipment operating at a high level of accuracy and precision ensures that the uncertainty of air density corrections is a very small fraction of the total uncertainty and is unlikely to have any impact when comparing results from different institutions.

The following instruments can be used:

(a)   Temperature: An accuracy in the measurement of temperature of 0.2°C is recommended (resulting in an uncertainty due to k_T of less than 0.1%). Miniature thermistors or platinum elements are readily available and provide this necessary accuracy. Given the environmental and health concerns with mercury-containing devices, mercury thermometers are not recommended.
(b)   Pressure: The recommended accuracy of any pressure sensor is 0.1 kPa, ensuring that the uncertainty due to k_p is less than 0.1%. Accurate digital barometers[2] are readily available, negating the need for mercury barometers. For the measurement of air pressure, it is not necessary to have the barometer located in the room where the source measurements are taking place, as

---

[2] Note that some smartphones contain an uncalibrated pressure sensor that may be useful to check a calibrated barometer. Anecdotal evidence indicates performance at the 0.1% level is possible. However, the lack of any calibration means such a device will only be used as a check. The user has to verify performance before using them for any secondary QA check.

long as it is close by and, ideally at nearly the same altitude. The pressure equation ($p = \rho g h$) can be used for small altitude corrections (e.g. from one floor of a building to another).

(c)   Relative humidity: In general, one does not need a very accurate measurement of humidity and the recommended accuracy is 5% *RH*. In general, humidity values are more often used for a go/no-go decision (i.e. to ensure that the relative humidity of the air in the room where the ionization chamber will be used lies between acceptable limits (20% to 70% *RH*)). Any suitable and calibratable device can be used for this measurement. The measuring range of the humidity meter is expected to cover at least 15% to 80% *RH*.

According to the ISO/IEC17025 [124] standard, all equipment that impacts a calibration will have a traceable calibration. In the context of brachytherapy dosimetry, this is generally followed for the well-type chamber and electrometer, but it is not always followed for associated sensors such as temperature and pressure meters. This code of practice recommends thermometers and barometers to have regular calibrations which are traceable to primary standards. How often the recalibration needs to be performed is determined by the experience and depends on the type of instrument used [125]. The calibration interval can be wider for hygrometers [88]. As general guidance, a time interval for recalibration less than two years is not likely to be necessary, but annual checks are recommended (e.g. by comparison with secondary check instruments or through the exchange with a sister institution).

5. DOSIMETRY FRAMEWORK

The need for accurate radiation dose measurements is highlighted in the medical use of radiation, particularly in radiation therapy. Regular calibration of dosimetry equipment has been a common practice in EBRT [40, 126]. The situation has been different in brachytherapy, where calibrations have not been as prevalent in the past. Manufacturers supply the brachytherapy sources with a source certificate and in many cases, this has been used directly for the patient treatment. However, this is not considered to be appropriate, and an independent verification is needed. In patient dosimetry the aim is to achieve an adequate level of accuracy, with the uncertainty of the source strength being less than 3% ($k = 2$) [18, 19]. This is achievable only if calibrated dosimeters are used for measurements.

5.1. CLASSIFICATION OF INSTRUMENTS AND STANDARDS

Instruments can be classified according to the following categories, which are adapted from the list provided in [127]:

(a) Primary standard: An instrument of the highest metrological quality allowing the determination of a unit of a given physical quantity and not referring to other standards of the same quantity. The accuracy has been verified by comparison with standards for the same quantity maintained by other institutes participating in the International Measurement System (IMS).

(b) Secondary standard: An instrument with established precision and long term stability that has a calibration traceable to a primary standard for the same physical quantity to be measured.

(c) National standard: A standard recognized by a national authority as the basis for assigning the value in a country of all other standards of the given quantity.

(d) Working standard: An instrument having the highest metrological quality available at a given location, which is used routinely to calibrate measuring equipment. Especially since the measurements by primary standards are complicated and time-consuming, they are not always feasible for routine calibrations of brachytherapy sources and chambers at the PSDLs. Therefore, working standards are applied also at the PSDL level for routine calibrations [36].

5.2. THE INTERNATIONAL MEASUREMENT SYSTEM

According to the Safety Standards Series No. GSR Part 3, Radiation Protection and Safety of Radiation Sources: International Basic Safety Standards [42], the calibration of all dosimeters used for patient dosimetry and those for source calibration should be traceable to a standards dosimetry laboratory. The IMS is the technical and administrative infrastructure ensuring that measurements can be performed at an accuracy that is fit for purpose [128]. The standards used for calibrations should be traceable to the International System of Units (SI) and they may be either secondary or primary standards. This international arrangement for traceability is represented schematically in Fig. 4. If a country does not have a national reference standard, they need to arrange access to such standards in another country.

Figure 4 is indicative of the dissemination of the traceability for a brachytherapy standard starting from PSDLs which provide calibrations to SSDLs

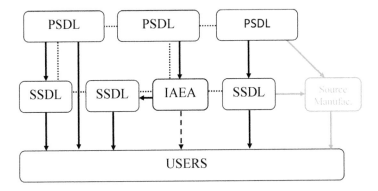

FIG. 4. A simplified representation of the IMS for HDR brachytherapy dosimetry. The solid black arrows represent the typical calibration chains, and the dotted lines indicate comparisons of primary and secondary standards. The dashed arrow is indicating an exceptional calibration of a user instrument by the IAEA that might happen if a country has very limited resources and no calibrations available for brachytherapy. The grey arrows represent the source calibration provided by the manufacturer.

or manufacturers. The Bureau International des Poids et Mesures (BIPM) does not currently offer brachytherapy calibrations but their evaluation of reference air kerma rate is taken as the key comparison reference value. The PSDLs and SSDLs intercompare their standards periodically so that it is confirmed that they agree within their stated uncertainties. Users are advised to not directly rely on the dosimetry values from a manufacturer but to confirm that their measurements are traceable to a PSDL.

The process for the user applying the traceability varies whether LDR seeds are used, or an HDR/PDR source is used. The user employing an HDR afterloader source needs to have a traceable well-type ionization chamber calibration from an SSDL or a PSDL (Fig. 4). The RAKR (or AKS) value from the source certificate is only an indicative value for regulatory purposes. In the case of LDR seeds (Fig. 5) the usual clinical application is on the order of 100 seeds. The user is advised to measure 10% or at least 10 seeds with a calibrated well-type ionization chamber and compare the results with the manufacturer value. The value measured by the user and the value stated by the manufacturer are advised to agree within 5%; if this is not the case, the user needs to consult with the manufacturer and the radiation oncologist [25].

5.2.1. The role of PSDLs

PSDLs have an important role in the calibration chain. They establish the quantity and disseminate it to SSDLs and end users. PSDLs provide calibrations

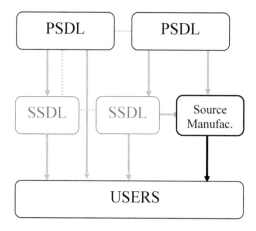

FIG. 5. A simplified representation of the IMS for LDR brachytherapy dosimetry. The grey arrows indicate the calibrations, and the dotted lines indicate comparisons of primary and secondary standards. The black arrow represents the typical source calibration chain provided by the source manufacturer.

for suitable dosimetry equipment, such as well-type chambers but also for radioactive sources. In 2020, there were 13 national metrology institutes or designated institutes (10 PSDLs and 3 SSDLs) which have brachytherapy calibration services in terms of RAKR listed in the key comparison database of BIPM [129]. A more detailed description of primary standards and their dissemination is given in the following section.

5.2.2. The role of SSDLs

There are only 10 PSDLs in the world providing RAKR (or S_K) calibrations and they cannot provide calibrations for all end users. The role of SSDLs is to bridge this gap and provide calibrations for end users. Typically, an SSDL will maintain that country's national reference standard and they provide calibrations within the country or in a region. In the United States of America (USA), the secondary standards dosimetry laboratories are called accredited dosimetry calibration laboratories (ADCL). In the USA, there are three ADCLs that provide brachytherapy calibration services for LDR and HDR applications. The ADCLs intercompare with NIST on a regular schedule. The ADCLs in the USA maintain sources of each radionuclide and a calibrated well-type chamber.

The IAEA has established a network of SSDLs in collaboration with the World Health Organization (WHO). The objective of the IAEA/WHO SSDL Network is to provide and maintain the links between the end users, SSDLs and the IMS. In addition, the aim is to achieve consistency of the measurement, and IAEA codes of practice are supporting this goal. In 2020, there were 86 members and 16 affiliated members in the SSDL Network [130].

Based on the annual report of IAEA/WHO SSDL Network members, there are 28 SSDLs, that offer brachytherapy calibration services. However, only four SSDLs actually provided calibrations in 2016. Based on a survey performed in 2017, most of them have only a calibrated well-type chamber and do not have any brachytherapy sources on-site. Therefore, cross-calibrations are often performed in hospitals.

The IAEA Dosimetry Laboratory is a central laboratory of the SSDL Network, and the IAEA provides calibration, comparisons, and audit services for the Members States. In 2020, the IAEA provided calibration services for reference standards using HDR ^{192}Ir and ^{60}Co sources and LDR ^{137}Cs sources. A current list of services can be found on the SSDL Network website [131].

5.2.3. Recognition of calibration services

In addition to maintaining an operational calibration service, the calibration laboratories will have appropriate quality systems and defined procedures for calibration work. The professionals working in the calibration laboratories (radiation metrologists) will have specific qualifications for brachytherapy calibrations [130]. The SSDL will demonstrate their capabilities by regularly participating in comparisons and audits, and by having external reviews to show supporting evidence for their calibration service.

The International Committee for Weights and Measures (CIPM) has established a mutual recognition arrangement [132, 133], which provides a route to get an international approval for the calibration and measurement capabilities. Those countries that belong to the Meter Convention, the intergovernmental organization which allows Member States to act in agreement on all matters related to units of measurement, can achieve international recognition within the CIPM mutual recognition arrangement. A dosimetry laboratory has to take part in relevant measurement comparisons and have their quality management systems (QMSs) established, reviewed and approved.

The IAEA SSDL Charter sets forth the minimum requirements for the SSDLs that want to become a member of the IAEA/WHO SSDL network. Also, in this case, the SSDLs have to provide traceable calibrations, estimate uncertainties, participate in comparisons and demonstrate the quality of their measurements through a quality management systems in line with the ISO/IEC 17025 standard [124].

The international standard ISO/IEC 17025 [124] covers general requirements with regard to the competence of testing and calibration laboratories.

The standard contains a set of requirements relevant to calibration laboratories to demonstrate that they are technically competent to perform the work that they do.

Comparisons are an important part of the IMS. However, only few key comparisons have been organized [134–138]. The IAEA provides comparison services for radiation therapy, radiation protection and diagnostic radiology but not yet for brachytherapy. Only one special technical event with a comparison exercise has been organized. SSDLs are encouraged to promote setting up RAKR or S_K comparisons and regional brachytherapy audits to ensure consistency in brachytherapy dosimetry across different radiotherapy centres.

5.2.4. Role of the manufacturer

In brachytherapy, the role of the manufacturer is to provide radioactive sources that are consistent in their strengths as ordered by the user. A source certificate is also provided in support of the radioactive sources (see Section 8.6). For HDR applications, the manufacturer will supply a source with contained activity close to that ordered by the user, but it is the user's responsibility to measure its output and provide the RAKR or S_K for input to the TPS. Manufacturers of the radioactive sources have a different role for LDR sources. Since large numbers of sources are used and the user may not measure all of them, the manufacturer will supply sources with uniform output. All other aspects to help the user will be considered by the manufacturer. In all cases there is a need for manufacturers to establish and then maintain traceability to the international metrological network for all brachytherapy sources. An example of confirmation of traceability may be an annual procedure for LDR source suppliers, similar to the system used in North America [88]. In all cases, the possibility for errors exists if there is no independent end user verification of the RAKR or S_K, which is advised to be performed with a traceably calibrated instrument. Manufacturers are also advised to have a traceably calibrated instrument to maintain their consistency.

6. ESTABLISHMENT AND DISSEMINATION OF CALIBRATION QUANTITIES

In the following section information about the establishment of the recommended quantities is provided. Since their dissemination takes place mainly through the use of well-type chambers, specific information about the calibration of the well-type chamber dosimetry system is given. Request for and frequency of calibration as well as information provided in the calibration certificate are also discussed.

6.1. ESTABLISHMENT OF PRIMARY CALIBRATION STANDARDS

Since the publication of the IAEA-TECDOC-1274, Calibration of Photon and Beta Ray Sources Used in Brachytherapy: Guidelines on Standardized Procedures at Secondary Standards Dosimetry Laboratories (SSDLs) and Hospitals [36] in 2002, many national metrology institutes (NMIs) have developed and established new brachytherapy primary standards for the realization of the quantities reference air kerma rate, air kerma strength and absorbed dose to water. These instruments at the top of the calibration chain are of the highest metrological quality. For brachytherapy applications, they are used to calibrate secondary standard dosimeters, for instance well-type ionization chambers. The calibration chain and the dissemination of the physical units from PSDLs via SSDLs to the end users has been described in Section 5. The use of traceably calibrated dosimeters for the measurement of the source strength of brachytherapy sources by the end users (radiotherapy centres and source manufacturers) enables an accurate delivery of brachytherapy applications in the clinic.

Different types of primary standard instruments have been built or are currently being developed to enable the measurement of sealed LDR and HDR brachytherapy gamma sources, electronic brachytherapy miniature X-ray sources and ophthalmic applicators (beta sources). The variety of different design concepts for the brachytherapy primary standards that are currently available provides a robust system for the characterization of brachytherapy sources around the world.

Some of these devices have been compared to each other via international key comparisons organized by the BIPM over the last decade. At this stage, only comparisons for the measurement of reference air kerma rate for HDR ^{192}Ir brachytherapy sources have been established: BIPM.RI(I)-K8, APMP. RI(I)-K8 and EURAMET.RI(I)-S8. The results of these ongoing key and supplementary comparisons are being published in the BIPM key comparison database (KCDB) [139].

Appendix II provides a detailed overview of the primary standard measurement techniques that are currently in use at NMIs, ranging from ionometric to calorimetric and chemical dosimetry standards. For further technical details on specific instruments, readers are encouraged to refer to the scientific publications in the list of references.

6.2. CALIBRATION OF THE WELL-TYPE CHAMBER DOSIMETRY SYSTEM

The recommended method to measure the source strength of the main brachytherapy sources is based on the use of a calibrated well-type chamber system. The electrometer can be calibrated as a system with the well-type chamber, or separately. In a system calibration, the well-type chamber connected to the electrometer is calibrated at the same time, as a system; while in a component calibration, the chamber and electrometer are calibrated separately and the overall calibration can then be derived from the chamber calibration and the electrometer calibration.

Extension cables can be permanently fixed in place and therefore it is not always possible to include them in the complete system that is sent for calibration. In such cases, additional means could be used to check the integrity or leakage of any chamber extension cable, by allowing the electrometer to be connected directly to the chamber, and checks are advised to be repeated as part of a regular QA procedure (e.g. annually). This QA procedure can also be accomplished by a redundancy procedure [140], meaning a comparison of at least two well-type chambers.

Even if well-type chamber dosimetry systems have a proven long term stability [36, 38, 112], it is recommended to perform regular recalibration (e.g. every two or three years) [25, 37, 88, 113, 141–143], or in case of drifts in the chamber response indicated by a long term stability check (Section 8.5). This recommendation is in agreement with requirements for EBRT equipment [40]. Recalibration is also advised to be performed whenever the user suspects any damage of one component of the dosimetry system, or after any repair of one component of the dosimetry system that might possibly have changed its performance.

6.2.1. SSDL well-type chamber dosimetry system calibration

The SSDL's well-type chamber dosimetry system (i.e. secondary standard) should be calibrated against a suitable primary standard at a PSDL. If the SSDL does not have easy access to a PSDL, their well-type chamber can also be calibrated in another SSDL, ADCL or the IAEA Dosimetry Laboratory, provided that this laboratory is traceable to a PSDL. The well-type chamber calibration will ideally be performed for each source model, *sm*, which is expected to be used by the SSDL for subsequent calibrations of the end users' well-type chambers. However, in practice this is not possible since traceability is normally available only for a specific source, and therefore a source model correction factor needs to be applied. The specific source holder for the well-type chamber and the source model used needs to be chosen accordingly.

The well-type chamber calibration at a PSDL is a two-step process. Initially, the source strength of the brachytherapy source installed in the radiation facility at the PSDL is measured from first principles with a primary standard. The measured source is then used to calibrate secondary standard well-type chambers.

In the extraordinary case that it is not feasible for SSDLs to send their well-type chambers to PSDLs for regular recalibrations covering the whole range of different radionuclides, traceability can be maintained by checking the well-type chamber calibration with long-lived calibration sources (e.g. ^{241}Am or ^{137}Cs; see Section 8.5.1) and implementing a proper quality control procedure. An ^{241}Am source is needed to cover the energy region for low energy photons for checking the constancy of a well-type chamber, whereas a ^{137}Cs source is needed to cover the energy region for high energy photons. If the chamber response for each of the two check sources changes within less than 0.5%, it can be assumed that the well-type chamber calibration coefficients for the relevant high energy and low energy sources have remained constant, and the calibration interval may be extended to a maximum of six years. However, best practice would still be to get the well-type chamber recalibrated every two or three years, as mentioned in Section 6.2.

The same approach applies if a redundant measurement system with two or more well-type chambers is implemented (see Section 8.5.1).

6.2.2. Hospital well-type chamber dosimetry system calibration

A minimum of one HDR source should be available to the SSDL to provide well-type chamber calibrations. They can decide to purchase their own remote afterloader with a specific source model or perform calibrations by using an HDR afterloader based at a hospital. For the latter case, additional QA steps may be required prior to any calibration to verify afterloader performance. In both cases, the source model correction factor k_{sm,sm_0} is advised to be applied if the actual source model differs from the one used to calibrate the SSDL's well-type chamber.

All measurements and calibrations with HDR and PDR ^{192}Ir sources at SSDLs are advised to be performed within a reasonable time frame (less than one year) after receipt of a new source of activity ~0.4 TBq from the source manufacturer to achieve dose rates that are similar to those typically measured in clinics. Exchanges of HDR ^{60}Co sources can take place less frequently, preferably within a time frame of less than 10 years.

A selection of all types of low energy LDR brachytherapy seeds, which are expected to be used by the SSDL's end users will be available at the SSDL.

Alternatively, the end users could purchase their own low energy LDR sources and send them to the SSDL for calibration, provided the well-type chamber at the SSDL has been calibrated for that source type. The lowest calibration uncertainty can generally be obtained for sources listed in a recognized registry[3] as these types undergo regular evaluation for reference air kerma rate, anisotropy, and energy spectrum consistency. These seed types also benefit from published consensus data sets for the determination of absorbed dose to water. For other seed types, without such an evaluation program, anisotropy and spectrometry measurements are needed for full characterization, otherwise significantly higher uncertainties need to be assigned.

With both HDR and LDR calibration sources, source strength is advised to be within the same order of magnitude as the source strength of the brachytherapy sources used by the end user. Calibration coefficients of some well-type chambers, as for instance the PTW 33004/Nucletron SDS 077.09x, have shown dependency on the source activity [115].

Electrometers are typically calibrated at multiple points, having both low and high settings for collected charge and current. This approach allows their use for LDR sources, HDR sources, and EBRT dose measurements. Be aware of the leakage of the electrometer. Some older electrometers have higher leakage that can interfere with the measurement of LDR sources.

6.3. REQUEST FOR DOSIMETRY SYSTEM CALIBRATION

In order to initiate the calibration, some important information has to be provided from the clinic to the SSDL, or from the SSDL to the PSDL:

(a) The name and address of the institute.
(b) The contact information (name, email, phone).
(c) Type of calibration requested:
    — calibration quantity;
    — the radionuclide;
    — source model (optional).
(d) A list of all parts that are shipped (e.g. well-type chamber, source holder, electrometer, extension cable and their serial numbers).

It is recommended to adequately pack all components so that they do not to get damaged during transportation. Prior to shipping between the clinic and the

---

[3] For example, http://rpc.mdanderson.org/rpc/BrachySeeds/Source_Registry.htm

SSDL, or the SSDL and the PSDL, and after receiving the shipment back, it is important to perform the following measurements:

(a) Stability check according to one of the methods given in Section 8.5. This will ensure that transportation did not affect the dosimeter response [40].

(b) Radioactive contamination check of the source holder of the well-type chamber. The local procedures for contamination verification are advised to be used. In HDR brachytherapy, since the entire source path (i.e. afterloader, transfer tube, source holder) is closed towards the outer environment and the source is not in communication with the external environment, it is not necessary to check contamination of the well-type chamber and the electrometer unless there are specific concerns for cross-contamination.

6.4. INFORMATION PROVIDED IN THE CALIBRATION CERTIFICATE OF THE DOSIMETRY SYSTEM

The following information needs to be provided on the calibration certificate of the dosimetry system:

(a) Details of the user's well-type chamber:
— Name of manufacturer;
— Model;
— Serial number.

(b) Details of the user's electrometer (if included for calibration):
— Name of manufacturer;
— Model;
— Serial number;
— Measurement range.

(c) Details of the source holder(s):
— Name of manufacturer;
— Model;
— Serial number/identifier;
— Type of any ancillary equipment (e.g. plastic catheters or steel needles, if applicable) to be used with the source holder.

(d) Date of calibration.

(e) Details of the brachytherapy source(s) used for calibration:
— Radionuclide;
— Name of manufacturer;
— Model;
— Serial number;

— RAKR or AKR and date of its measurement;

— Source model correction factors (see Sections 7.2 and 8.3).

(f) A brief description of the calibration procedure and experimental set-up. Traceability to the PSDL standard and possible application of the source model correction factor.

(g) Details on the methods and reference conditions used for calibration:

— The date of calibration;

— The standard polarizing voltage and the measurement range applied to the well-type chamber; the polarity of the central collecting electrode and guard electrode with respect to the outer electrode;

— The position of the source in the well-type chamber (i.e. sweet spot or fixed position);

— The reference environmental conditions for:

— Temperature T_0 (typically 293.15 K or 295.15 K);

— Air pressure P_0 (typically 101.325 kPa);

— Relative humidity RH_0 (typically 50%).

— The current environmental conditions for temperature T, air pressure P and relative humidity RH.

(h) Details on the calibration results:

— Information if a system or component calibration is performed;

— In case of system calibration: calibration coefficient of the well-type chamber $N_{\dot{K}_{\delta,R},sm_0}$ (or N_{S_K,sm_0} or $N_{\dot{D}_w,sm_0}$), including the uncertainty and its confidence level. The electrometer calibration coefficient k_{elec} equals 1 in this circumstance;

— In case of component calibration: the calibration coefficient of the well-type chamber $N_{\dot{K}_{\delta,R},sm_0}$ (or N_{S_K,sm_0} or $N_{\dot{D}_w,sm_0}$) and of the electrometer k_{elec} given separately, including uncertainties and their confidence levels.

The measurement unit for $N_{\dot{K}_{\delta,R},sm_0}$ or $N_{\dot{D}_w,sm_0}$ is usually Gy h^{-1} A^{-1}, whereas the measurement unit for N_{S_K,sm_0} is usually Gy h^{-1} A^{-1} m^2. The uncertainty in the calibration coefficient will be reported as an expanded uncertainty with a coverage factor $k = 2$ providing a coverage probability of approximately 95% [142].

7. DOSIMETRY FORMALISM

The formalism employed for the determination of dosimetry quantities used in brachytherapy is similar to the formalism used in TRS-398 [40] for the determination of absorbed dose to water in external beam radiotherapy and in TRS-457 [41] for the determination of air kerma in diagnostic radiology. The formalism given in this code of practice is based on standards of air kerma rate (i.e. S_K and $\dot{K}_{\delta,R}$) and is valid for the radioactive photon-emitting brachytherapy sources considered in this code of practice. An exception is given by beta sources, where the recommended calibration quantity is based on absorbed dose to water standards. With regard to beta sources, the current code of practice applies only to IVBT sources and excludes ophthalmic applicators and plaques. Since in some countries calibrations of photon sources based on primary standards in terms of absorbed dose to water have been established, even if their availability is currently limited [76–78], the formalism based on $\dot{D}_W$ is also provided at the end of this section.

7.1. FORMALISM BASED ON STANDARDS OF REFERENCE AIR KERMA RATE

The air kerma rate at the time t, $\dot{K}(t)$, produced at the reference point by the radiation emitted by a reference source model sm_0 and in the absence of the dosimeter, is given by:

$$\dot{K}(t) = N_{\dot{K},sm_0} M_{sm_0}(t) \tag{5}$$

where $M_{sm_0}(t)$ is the reading of the dosimeter at time t under the reference conditions used in the standards laboratory and corrected for the influence quantities, and $N_{\dot{K},sm_0}$ is the calibration coefficient of the dosimeter in terms of air kerma rate obtained from a standards laboratory under reference conditions of irradiation. In Eq. (5), $\dot{K}$ represents a generic air kerma rate term for one of the dosimetric quantities S_K and $\dot{K}_{\delta,R}$ used in this code of practice and defined previously.

7.1.1. Reference conditions

Reference conditions are given by a set of values of influence quantities for which the calibration coefficient $N_{\dot{K},sm_0}$ is valid without the need for further corrections. Reference conditions for calibrations in terms of air kerma rate that have to be considered when performing dosimetry in brachytherapy are, for instance, those that involve the ambient temperature, pressure and relative humidity, the radiation quality emitted by the source, the geometrical arrangement of the source with respect to the detector, etc. Since measurement conditions are usually different from conditions in the standards laboratories, additional corrections need to be considered.

7.1.2. Influence quantities

Reference conditions are defined by a set of values of influence quantities for which the calibration coefficient is valid without any further correction. Influence quantities are defined "as quantities not being the subject of the measurement, but yet influencing the quantity under measurement" [40]. They may be of different nature as, for example, ambient pressure, temperature, and relative humidity; they may arise from the dosimetry instrumentation (e.g. leakage, polarization, ion recombination), or may be quantities related to the radiation quality (e.g. source model).

As many influence quantities as is practicable are kept under control during the measurement. However, many influence quantities such as air pressure or dose rates of radioactive sources cannot be controlled. Appropriate correction factors are thus established to take into account the effect of these influence quantities. Under the assumption that influence quantities act independently from each other, a product of correction factors k_i can be applied to the raw reading $M_{sm_0,\text{raw}}(t)$ according to:

$$M_{sm_0}(t) = M_{sm_0,\text{raw}}(t) \prod_i k_i \tag{6}$$

where each k_i is related to the ith influence quantity only and $M_{sm_0}(t)$ is the corrected measurement.

In analogy to TRS-398 [40], a departure from the reference radiation quality used to calibrate the dosimetric system was treated as an influence quantity and not included among the correction factors k_i above. Measurements with a source model sm other than the source model used for calibration, sm_0, are therefore treated explicitly by the *source model correction factor* k_{sm,sm_0}, which is described in detail below.

7.2. SOURCE MODEL CORRECTION FACTOR

A calibration source of one radionuclide and encapsulation may generate a well-type chamber response that differs from that of another source with the same source strength but of different model, even if the radionuclide is the same [141, 143, 144]. Higher discrepancies are expected if different radionuclides are selected. Source model correction factors (k_{sm,sm_0}) are therefore introduced to correct for differences between the radioactive source model sm_0 used at the calibration laboratory and the actual source model sm used by the end user. They also depend on the type, model, and year of manufacture of the well-type chamber.

When the well-type chamber is used with a source model sm that differs from the one used for calibration, sm_0, the source strength of a photon emitting source sm at the time of measurement, t, can be determined as:

$$\dot{K}(t) = N_{\dot{K},sm} M_{sm}(t) k_{sm,sm_0} \tag{7}$$

where $N_{\dot{K},sm_0}$ is the calibration coefficient for the well-type chamber obtained with the reference source model sm_0. $M_{sm}(t)$ is the reading of the dosimeter at the time t, performed with a source model sm and corrected for all influence quantities, and k_{sm,sm_0} is the source model correction factor that takes into account differences between sm and sm_0. Values of k_{sm,sm_0} to be used in Eq. (7) apply only to air kerma rate measurements. As in Eq. (5), $\dot{K}(t)$ represents a generic air kerma rate term at the time t, for one of the reference dosimetric quantities S_K and $\dot{K}_{\delta,R}$ used in this code of practice.

In some specific cases, k_{sm,sm_0} is equal to the 'source geometry factor' (k_{sg}) defined recently in the Institute of Physics and Engineering in Medicine (IPEM) Code of Practice for HDR brachytherapy [141], but also extends the definition to sources different from ^{192}Ir. In Schüller et al. (2015) [145] the chamber type-specific correction factor was named 'radiation quality correction factor' k_Q and was introduced to enable the measurement of the RAKR of an HDR ^{60}Co source by means of a well-type chamber calibrated with an HDR ^{192}Ir source. However, at this stage this code of practice does not recommend corrections between different radionuclides, as consensus data are not available to indicate that such a conversion can be achieved with the necessary accuracy.

Ideally, the well-type chamber is advised to be calibrated with the same source model that is used in the clinic for treatment. Since in that case sm would be equal to sm_0, k_{sm,sm_0} would be unity. However, it is reasonable that national dosimetric standards are usually based on one single source model [144], whereas hospitals of the same country may use different source models in their clinical practice. If no consensus data are available and sm and sm_0 are constituted by the same radionuclide but different models, k_{sm,sm_0} should be taken to be unity until consensus data become available. In this case, the user should apply an additional uncertainty to the calibration coefficient[4] of the well-type chamber if it is used for the measurement of source types that are different from sm_0.

7.3. SOURCE DECAY CORRECTION FACTOR

As the radionuclide source strength changes with time due to radioactive decay, it is necessary to correct the source strength at the time of measurement t_{meas} to the source strength at some reference time t_{ref}. This happens when comparing the measured source strength to the value listed on the calibration certificate (where t_{ref} generally occurs before t_{meas}) or when the source strength is used for a patient dosimetry calculation (where t_{ref} occurs after t_{meas}). The source decay correction factor k_{dec} is defined as:

$$k_{dec} = e^{(t_{meas} - t_{ref})\left(\frac{\ln 2}{t_{1/2}}\right)} = 2^{\frac{t_{meas} - t_{ref}}{t_{1/2}}} \tag{8}$$

where $t_{1/2}$ is the radionuclide half-life. Note that k_{dec} equals unity when $t_{ref} = t_{meas}$. k_{dec} is applied to correct the source strength at the time of measurement according to:

$$\dot{K}(t_{ref}) = \dot{K}(t_{meas})k_{dec} \tag{9}$$

Taking into account Eq. (9), Eq. (7) becomes:

$$\dot{K}(t_{ref}) = N_{\dot{K},sm_0} M_{sm}(t_{meas})k_{dec}k_{sm,sm_0} \tag{10}$$

Recommended half-lives for some of the radionuclides used in brachytherapy are provided in Table 3. Care should be taken with the selection of the same temporal units for the time difference and the half-life. For unit conversion of the half-life from years (y) to days (d), the factor 365.242198 d/y that takes into account leap years needs to be considered. Users are also recommended to ensure that the same time standard is chosen for both t_{ref}, and t_{meas}, accounting for differing time zone and summer time corrections as well as for potentially differing formats for expressing date and time [104].

---

[4] Note that it is not possible to give any general indication on the size of this extra uncertainty because of the variety of different brachytherapy sources which are currently available or might become available in the future.

7.4. FORMALISM BASED ON STANDARDS OF ABSORBED DOSE RATE TO WATER

The absorbed dose rate to water at the time t_{ref}, $\dot{D}_{W,R}(t_{ref})$ produced at the reference point $P(r_0, \theta_0)$ by the radiation emitted by a reference source model sm, which is calibrated to absorbed dose to water and in the absence of the dosimeter, is given by:

$$\dot{D}_{W,R}(t_{ref}) = N_{\dot{D}_W, sm_0} M_{sm}(t_{meas}) k_{dec} k_{\dot{D}_W, sm, sm_0} \tag{11}$$

where M_{sm} is the dosimeter reading under the reference conditions used in the standards laboratory, $N_{\dot{D}_W, sm_0}$ is the calibration coefficient of the dosimeter in terms of absorbed dose rate to water, obtained from a standards laboratory under reference conditions of irradiation, and $k_{\dot{D}_W, sm, sm_0}$ is the source model correction factor specific for the $N_{\dot{D}_W, sm_0}$ based formalism.

All the considerations with regard to reference conditions, influence quantities, source decay and cross-calibrations provided in the Sections 7.1, 7.3 and 7.7 apply in the same way to the $N_{\dot{D}_W, sm_0}$ based formalism. Equation (6) has to be applied to correct for influence quantities other than the radiation quality. Due to the lack of available data, the use of any source model correction factor $k_{\dot{D}_W, sm, sm_0}$ is not recommended at this stage.

7.5. DETERMINATION OF THE REFERENCE SOURCE STRENGTH

For reference dosimetry of a brachytherapy source, it is assumed that a well-type chamber with known calibration coefficient traceable to a primary standard is available. The calibration coefficient is provided under reference conditions for a reference source model sm_0.

According to Eq. (10), which is valid for all the radioactive brachytherapy sources to which this code of practice applies, the RAKR $\dot{K}_{\delta,R}(t_{ref})$ and the AKS $S_K(t_{ref})$ of the actual radioactive source sm at the reference time t_{ref} can be determined according to:

$$\dot{K}_{\delta,R}(t_{ref}) = N_{\dot{K}_{\delta,R}, sm_0} M_{sm}(t_{meas}) k_{dec} k_{sm, sm_0} \tag{12}$$

and

$$S_K\left(t_{ref}\right) = N_{S_K,sm_0} M_{sm}\left(t_{meas}\right) k_{dec} k_{sm,sm_0} \tag{13}$$

respectively. $M_{sm}(t_{meas})$ is the reading of the dosimeter at the time t_{meas}, corrected for the source decay k_{dec} and influence quantities excluding the source model correction factor k_{sm,sm_0}, which is treated separately. $N_{\dot{K}_{\delta,R},sm_0}$ and N_{S_K,sm_0} are the RAKR and AKS calibration coefficients, respectively.

In case the $N_{\dot{D}_W,sm_0}$ based formalism is used, according to Eq. (11), the absorbed dose rate to water at the reference point $P(r_0, \theta_0)$ given by the actual radioactive source sm at the reference time t_{ref} can be determined as:

$$\dot{D}_{W,R}\left(t_{ref}\right) = N_{\dot{D}_W,sm_0} M_{sm}\left(t_{meas}\right) k_{dec} k_{\dot{D}_W,sm,sm_0} \tag{14}$$

where $N_{\dot{D}_W,sm_0}$ is the calibration coefficient of the dosimeter in terms of absorbed dose rate to water obtained from a standards laboratory under reference conditions of irradiation and M_{sm} is the reading of the dosimeter at the time t_{meas}, corrected for the source decay k_{dec}, the source model $k_{\dot{D}_W,sm,sm_0}$ and the influence quantities.

7.6. CALIBRATION OF THE WELL-TYPE CHAMBER DOSIMETRY SYSTEM

For the calibration of the user well-type chamber dosimetry system at the dosimetry laboratory, it is assumed that the value of the reference dosimetry quantity for a source model sm_0 (i.e. $\dot{K}_{\delta,R}, S_K$ or $\dot{D}_{W,R}$), measured under reference conditions, is known. The source strength of sm_0 can be measured at the SSDL level with a calibrated well-type chamber (traceable to a primary standard) according to the principles provided in Section 7.5 and the procedure provided in Section 8. Otherwise, some PSDLs also offer a brachytherapy source calibration service. Sources are calibrated against the PSDL's primary standard, and the traceably calibrated sources can then be shipped to SSDLs for the subsequent calibration of either the SSDLs' own well-type chambers or the end users' well-type chambers.

According to Eq. (12) and Eq. (13), the RAKR calibration coefficient $N_{\dot{K}_{\delta,R},sm_0}$ and the AKS calibration coefficient N_{S_K,sm_0} can be determined with:

$$N_{\dot{K}_{\delta,R},sm_0} = \frac{\dot{K}_{\delta,R}\left(t_{ref}\right)}{M_{sm_0}\left(t_{meas}\right) k_{dec}} \tag{15}$$

and

$$N_{S_{\mathrm{K}},sm_0} = \frac{S_{\mathrm{K}}\left(t_{\mathrm{ref}}\right)}{M_{sm_0}\left(t_{\mathrm{meas}}\right)k_{\mathrm{dec}}} \tag{16}$$

respectively. $\dot{K}_{\delta,\mathrm{R}}\left(t_{\mathrm{ref}}\right)$ and $S_{\mathrm{K}}\left(t_{\mathrm{ref}}\right)$ are the known reference dosimetry quantities RAKR and AKS at the reference time t_{ref} for the source model sm_0, respectively. $M_{sm_0}\left(t_{\mathrm{meas}}\right)$ is the reading of the dosimeter at the time t_{meas}, corrected for the influence quantities (see Section 7.1.2) in order to fit the reference conditions for which the calibration coefficient will be valid. To get the hypothetical readings that would result at the reference time t_{ref}, the source decay correction factor is applied.

If the $N_{\dot{D}_{\mathrm{W}},sm_0}$ based formalism is used, according to Eq. (11) and in analogy to Eq. (15) and Eq. (16), the absorbed dose rate to water calibration coefficient $N_{\dot{D}_{\mathrm{W}},sm_0}$ is given by:

$$N_{\dot{D}_{\mathrm{W}},sm_0} = \frac{\dot{D}_{\mathrm{W},\mathrm{R}}\left(t_{\mathrm{ref}}\right)}{M_{sm_0}\left(t_{\mathrm{meas}}\right)k_{\mathrm{dec}}} \tag{17}$$

where $\dot{D}_{\mathrm{W},\mathrm{R}}\left(t_{\mathrm{ref}}\right)$ is the known reference absorbed dose to water at the reference time t_{ref} for the source model sm_0.

7.7. CROSS-CALIBRATION OF THE WELL-TYPE CHAMBER DOSIMETRY SYSTEMS

Traceability of reference dosimetry is obtained through the use of reference ionization chambers calibrated on a regular basis at a standards laboratory. While it is not desirable, or not practicable, to use a reference chamber in all clinical situations and for all routine measurements, any substitute field chamber that is used for this purpose is also required to have a calibration traceable to a national standard. This is achieved in the clinic through a process called cross-calibration, in which the calibration coefficient of the reference chamber for a specific source type is used to determine the required calibration coefficient of the field chamber. Because of the energy dependence of well-type chamber calibration coefficients, the same source type (i.e. same radionuclide) as used in the calibration of the reference instrument has to be used in the cross-calibration procedure. Depending on the particular situation, the same source holder may be used for both chambers, or a chamber-specific source holder used for each chamber to allow complete separation of apparatus. In either case, the source holder as used in the primary calibration has to be used for the reference chamber. Any alternative approach using, for example, a different source type (i.e. different radionuclide) combined with interpolation and/or calculated correction factors is not permitted.

The reference chamber has a calibration coefficient $N_{\dot{K},sm_0}^{\mathrm{ref}}$ (i.e. $N_{\dot{K}_{\delta,R},sm_0}^{\mathrm{ref}}$ or $N_{S_K,sm_0}^{\mathrm{ref}}$) for source model sm_0. If the user of the traceably calibrated chamber (either an SSDL or a hospital) has access to the same source model sm_0 as used for calibration, this source model should also be used for the cross-calibration of the field chamber. If the user has only access to a different source model sm, the calibration coefficient of the reference chamber has to be multiplied by an appropriate source model correction factor (unity if $sm = sm_0$ in Eqs (18), (19) and (20)).

Source model sm is positioned in the well-type chamber using the same source holder as for when the calibration was obtained, at the same dwell position within the well. A measurement is obtained, M_{sm}^{ref}, (current or charge for a fixed time), corrected for influence quantities. The same measurement is then obtained for the field chamber, yielding M_{sm}^{field}, also correcting for influence quantities. Both measurements are recommended to refer to the same time t_{ref}. The source decay correction factor will eventually be used to correct the reading from t_{meas} to t_{ref}. Depending on the apparatus used (e.g. same or different electrometer, same or different source holder) additional checks and/or warm-up times may be required. Combining the two measurements with the known calibration coefficient of the reference chamber gives:

$$N_{\dot{K}_{\delta,R},sm}^{\mathrm{field}} = \frac{M_{sm}^{\mathrm{ref}}\left(t_{\mathrm{ref}}\right)}{M_{sm}^{\mathrm{field}}\left(t_{\mathrm{ref}}\right)} N_{\dot{K}_{\delta,R},sm_0}^{\mathrm{ref}}\, k_{sm,sm_0} \tag{18}$$

$$N_{S_K,sm}^{\mathrm{field}} = \frac{M_{sm}^{\mathrm{ref}}\left(t_{\mathrm{ref}}\right)}{M_{sm}^{\mathrm{field}}\left(t_{\mathrm{ref}}\right)} N_{S_K,sm_0}^{\mathrm{ref}}\, k_{sm,sm_0} \tag{19}$$

for $N_{\dot{K}_{\delta,R},sm_0}^{\mathrm{ref}}$ and $N_{S_K,sm_0}^{\mathrm{ref}}$ calibration coefficients, respectively.

The same considerations apply to calibration coefficients based on absorbed dose to water standards, leading to the equation:

$$N_{\dot{D}_W,sm}^{\mathrm{field}} = \frac{M_{sm}^{\mathrm{ref}}\left(t_{\mathrm{ref}}\right)}{M_{sm}^{\mathrm{field}}\left(t_{\mathrm{ref}}\right)} N_{\dot{D}_W,sm_0}^{\mathrm{ref}}\, k_{\dot{D}_W,sm,sm_0} \tag{20}$$

The calibration coefficient obtained for the field chamber (i.e. $N_{\dot{K}_{\delta,R},sm}^{\mathrm{field}}, N_{S_K,sm}^{\mathrm{field}}$ or $N_{\dot{D}_W,sm}^{\mathrm{field}}$) is applicable under the same reference conditions that existed during cross-calibration of the field chamber against the reference chamber, which, in turn, will be the same reference conditions that were used for the calibration of the reference chamber to derive its calibration coefficient at the calibration laboratory (i.e. $N_{\dot{K}_{\delta,R},sm_0}^{\mathrm{ref}}, N_{S_K,sm_0}^{\mathrm{ref}}$ or $N_{\dot{D}_W,sm_0}^{\mathrm{ref}}$).

8. CODE OF PRACTICE FOR WELL-TYPE CHAMBER CALIBRATION AND SOURCE STRENGTH MEASUREMENT

The calibration chain for photon-emitting brachytherapy sources is based on the use of similar equipment at each stage, from dissemination at the PSDL, through calibration at the SSDL to end use in a hospital. Therefore, similar measurement procedures are repeated at each stage. However, the procedures need to be harmonized to minimize uncertainties within the calibration chain.

This section is dedicated to providing guidance on the correct measurement procedure to obtain the optimal uncertainty in the calibration of well-type chamber dosimetry systems performed by standards laboratories and in the measurement of the strength of brachytherapy sources. A common measurement procedure is laid out below that can be used by SSDLs, clinical medical physicists and manufacturers with various photon-emitting radioactive sources – both high energy and low energy, HDR and LDR – and beta-emitting IVBT sources. Some of the best practice recommendations presented below are not unique to well-type chamber measurements and could apply to all ionization chamber measurements.

Since calibration of well-type chamber dosimetry systems and source strength measurement both rely on the measurement of the ionization current generated by a brachytherapy source inserted in the well-type chamber and corrected for the influence quantities, a common procedure is given below.

8.1. EXPERIMENTAL SET-UP AND EQUIPMENT PREPARATION

It is assumed that the source measurements are carried out in a room suitably shielded from any external radiation source that could significantly impact the calibration of the brachytherapy source in question. The well-type chamber and electrometer employed are advised to be reference class instruments satisfying the requirements given in Section 4.2. All HDR source measurements should be performed in a minimum scatter environment [118], with the chamber at a minimum distance of 1 m from any wall or floor [141]. For measurements of low energy LDR sources, the chamber distance from the wall or floor can be less than 1 m. However, the user is always advised to ensure that the contribution to the detector reading from scatter is less than 0.1% of the measured signal [146]. A low Z table/support should be used for the well-type chamber (e.g. plastic or wood <15 mm thickness) or on a thick yet stable foam support. In addition, any significant source of scatter within 1 m of the well-type chamber needs to be avoided. The length of the transfer tube will determine how far away the afterloader can be positioned from the chamber. The room is supposed to be air-conditioned for constancy of temperature and relative humidity (*RH*). Any variation of the air temperature is advised to be less than 0.5°C per hour and needs to be documented. Relative humidity is advised to be in the range 20% to 70%, where lower levels cause concern for static build up and higher levels cause concern for condensation.

It is important that the chamber and relevant source holders reach equilibrium with the ambient conditions before beginning calibration; at least 30 minutes is usually needed. Since significantly longer times can, however, be required, it is recommended to let the well-type chamber settle in the room overnight. ESTRO booklet No. 8 [118] reports that it takes around 400 minutes to eliminate a 4°C temperature difference between the ambient temperature and the temperature inside a Standard Imaging HDR 1000 Plus well-type chamber, for instance. If a measurement takes place at a third-party site, before proceeding with the measurements, the well-type chamber and electrometer need to be given enough time to equilibrate after transport.

The associated electrometer needs also to be switched on some time before measurements, to allow adequate stabilization. Although some devices achieve stabilization within minutes, it is recommended to wait at least 30 minutes prior to any measurements. The voltage gradient will determine the polarity of the charge being collected by the electrometer and it is important to ensure that the same polarity is used as stated in the calibration certificate. After the warm-up period, the electrometer should be zeroed as described in the manufacturer's recommendation and a leakage current measurement should be performed afterwards.

Air density and relative humidity of the sensitive volume of the well-type chamber needs to be measured and systematically checked for changes during the measurement procedure. In general, it is not practicable, nor desirable, to place a sensor inside the sensitive volume of the chamber, and therefore some measurement analogue is required. It is not recommended to simply monitor the air temperature of the room within which the chamber is placed. Placing a temperature sensing device either inside the well without the source holder being present, or in close proximity to the outside of the well-type chamber housing, will result in a good estimate of the air temperature inside the sensitive volume. A temperature sensor can be taped to the outside of the chamber housing to achieve stable temperature readings. The air pressure and relative humidity can be monitored within the room used for the measurement.

Before data are acquired, it is also important to ensure that the measurement system has stabilized once a source is inserted. It is important to establish the behaviour during stabilization of a well-type chamber for each type of source/seed for which the chamber will be used (i.e. HDR, LDR, IVBT).

8.2. WELL-TYPE CHAMBER MEASUREMENTS

The general formalism for source strength determination and for the calibration of the well-type chamber dosimetry system is given in Section 7. Using the $N_{\dot{D}_w,sm_0}$ based formalism, Eq. (14) and Eq. (17) are used.

The RAKR of a radioactive source model sm at the reference time t_{ref}, in the absence of the chamber, is given by:

$$\dot{K}_{\delta,R}\left(t_{ref}\right)=N_{\dot{K}_{\delta,R},sm_0}M_{sm}\left(t_{meas}\right)k_{dec}k_{sm,sm_0} \tag{21}$$

where $N_{\dot{K}_{\delta,R},sm_0}$ is the RAKR calibration coefficient for the reference radioactive source model sm_0 and $M_{sm}(t_{meas})$ is the reading of the dosimeter at the time t_{meas} with the source at the sweet spot of the well-type chamber, corrected for the influence quantities and excluding the source model correction factor k_{sm,sm_0}, which is treated separately, and k_{dec} is the correction factor for source decay. The AKS, for a source model sm at the reference time t_{ref}, in the absence of the chamber, is given by:

$$S_K\left(t_{ref}\right)=N_{S_K,sm_0}M_{sm}\left(t_{meas}\right)k_{dec}k_{sm,sm_0} \tag{22}$$

where N_{S_K,sm_0} is the AKS calibration coefficient for the reference radioactive source model sm_0 and $M_{sm}(t_{meas})$, k_{sm,sm_0} and k_{dec} are the same as described above.

For well-type chamber calibrations, the RAKR calibration coefficient can be determined according to:

$$N_{\dot{K}_{\delta,R},sm_0}=\frac{\dot{K}_{\delta,R}\left(t_{ref}\right)}{M_{sm_0}\left(t_{meas}\right)k_{dec}} \tag{23}$$

where $\dot{K}_{\delta,R}\left(t_{ref}\right)$ is the known RAKR for the source model sm_0 at the reference time t_{ref}, $M_{sm_0}\left(t_{meas}\right)$ is the reading of the dosimeter at the time t_{meas} with the source at the sweet spot of the well-type chamber, corrected for the influence quantities (see Section 7.1.2), and k_{dec} corrects for the source decay in the time range between t_{meas} and t_{ref}. The AKS calibration coefficient can be determined according to:

$$N_{S_K,sm_0} = \frac{S_K(t_{ref})}{M_{sm_0}(t_{meas})k_{dec}}$$ (24)

where S_K is the known AKS for the source model sm_0, with $M_{sm_0}(t_{meas})$ and k_{dec} being the same as described above.

Practical considerations for sweet spot determination, electrometer measurements and correction for influence quantities are given below in Sections 8.2.1, 8.2.2, 8.2.3 and 8.3, respectively. For completeness, some information on sweet spot determination and electrometer measurements is given also for IVBT sources. However, for IVBT source dosimetry, other guidelines can be consulted [26].

8.2.1. Sweet spot determination

The "calibration point of a well-type chamber is the point at which the centre of the source is positioned during the calibration procedure" [36]. For the highest accuracy of source strength determination, the calibration point is advised to be the position within the chamber where the signal is maximized (the sweet spot, see Section 4.2.1). Moreover, the source is expected to entirely fit within the sweet spot length of the chamber. Since the source position has a significant impact on the measurement, it is recommended to record the location of the calibration point on measurement worksheets.

To determine the sweet spot using an HDR source, it can be stepped through a series of vertical positions within the well of the chamber, either in the forward or backward direction depending on the stepper motor drive used by the afterloader. Dwell positions are not supposed to be separated by more than 2.5 mm from each other to ensure a reasonable regression of the response curve. The chosen limits for these measurement positions are advised to be at least at 10 mm distal and proximal to the expected sweet spot. Additional measurement positions with a broader separation, taken over a wider vertical range, can be included. Examples of relative well-type chamber responses for the reference-class well-type chambers listed in Table 5 are provided in Fig. 6. Other typical sweet spot determination results are shown in Refs [110, 141, 144, 145, 147].

It is important to note that the sweet spot location value depends on the adopted reference system used to define the source position inside the well-type chamber. It can, for instance, be measured relative to the bottom of the well, the bottom of the source holder, the first source dwell position inside the source holder, etc. For well-type chambers where a flexible plastic catheter or steel needle needs to be pushed into a universal source holder, small variations in the sweet spot position in terms of the dwell position displayed on the control unit of the afterloader can be expected between different measurement set-ups. The actual sweet spot of the well-type chamber would still be the same, but the corresponding displayed dwell position might differ depending on the length of the plastic catheter or how deep the catheter or steel needle is inserted into the source holder. For well-type chambers where a transfer tube with a given length can be connected to a fixed adapter at the top of the source holder this is less of an issue, and variations in the displayed dwell positions are typically well within ±1 mm between different measurement set-ups. However, even if the same transfer tube is used for consecutive measurements of the sweet spot, the displayed equivalent dwell position might still vary, for instance if the source drive mechanism needs to be adjusted during a planned maintenance of an afterloader. Best practice is therefore to perform a sweet spot measurement every time a source strength measurement is carried out.

An analytical procedure to locate the sweet spot can be provided for well-type chambers PTW 33005 Sourcecheck$^{4\pi}$, Standard Imaging HDR 1000 Plus and PTW 33004 (i.e. well-type chambers that show an increase and a decrease of the ionization current as a function of the source dwell position that can be approximated with a quadratic polynomial, if the source is positioned up to a few centimetres around the sweet spot). If the dwell position x_n (according to the adopted reference system) and the corresponding measured current $I(x_n)$ are annotated for the source up to 20 mm distal and proximal to the expected sweet spot, and the data is fitted with the quadratic polynomial equation:

$$I(x) = a\,x^2 + b\,x + c \tag{25}$$

The sweet spot x_{max} according to the adopted reference system can then be calculated as:

$$x_{max} = -\frac{b}{2a} \tag{26}$$

For the Standard Imaging IVB 1000 well-type chamber, Eq. (25) has not been used for determining the sweet spot location, since the well-type chamber axial response cannot be approximated as a quadratic polynomial. The sweet spot for this chamber is defined as the first local maximum of the well-type chamber response curve, as measured from the chamber entrance [141]. In Fig. 6, this is equivalent to the local maximum at the source dwell position of approximately 85 mm.

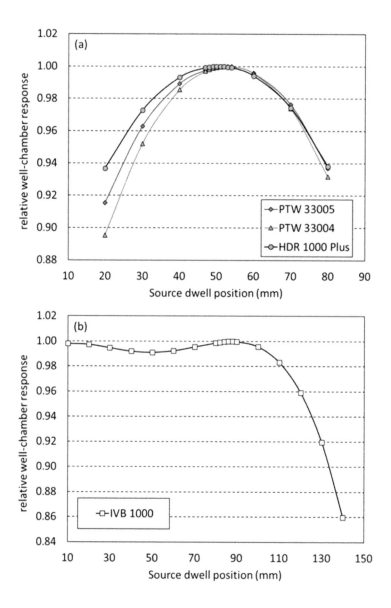

FIG. 6. Example of relative response curves, plotted with respect to the source dwell position (according to the adopted reference system) for an HDR ^{192}Ir source, for the well-type chambers (a) Standard Imaging 1000 Plus, PTW 33004 and PTW 33005 and (b) Standard Imaging IVB 1000.

In general, LDR source holders have a fixed geometry that is designed to position the radioactive seed at the calibration location. For those source holders that are not symmetric about their central long axis, it is advised to keep their rotational orientation within the well-type chamber constant. Whereas it is assumed that the chamber response will be independent of which end of the source is inserted, sweet spot determination is required only for source holders that may allow for height modifications.

8.2.2. Measurement techniques for current

Uncorrected measurements with the electrometer $M_{sm,\ raw}$ can be performed in two different ways:

(a) Measuring ionization current once the source has reached the point of measurement and the displayed ionization current has stabilized.

(b) Collecting the ionization charge for a specific time interval, starting and stopping it with the source stationary at the well-type chamber sweet spot (no source transit dose). Current is then obtained dividing the collected charge by the acquisition time.

In both cases, measurements with the source at the sweet spot of the well-type chamber should be performed. To avoid resolution errors due to the electrometer, it will likely be necessary to adjust the measurement time for charge measurements, depending on the source strength. Measurement time might be restricted due to restrictions imposed by the afterloader, with afterloaders often not allowing dwell times larger than 999 s. Measurements are always supposed to be performed with an electrometer setting that provides high resolution.

At both SSDLs and hospitals with photon emitting HDR sources, at least three source insertions to the chamber sweet spot will be made. For each source transfer, it is advised to perform a minimum of five measurements that are neither monotonically increasing nor decreasing (within 0.1%). The relative standard deviation of the mean is advised to be less than 0.1%, and the average of two sets of readings is advised to be within 0.2%. IPEM [141] reports that a 0.02% standard deviation is achievable for readings within a single source transfer into the chamber and therefore noted variations outside the 0.2% limits are advised to be investigated.

For photon emitting LDR source calibrations at the SSDL, at least three source insertions to the fixed measuring position will be made. At least three measurements of a single LDR seed are performed to check measurement reproducibility. After marking an end of the source, sensitivity of results to source orientation will be determined (typically less than a 0.4% difference). Sequential measurements in the same orientation are advised to be neither monotonically increasing nor decreasing beyond 0.1%. In the clinical setting, LDR source measurements do not require repeated source insertions if multiple sources from the same manufacturing lot are separately measured. At least three measurements of each seed are performed to check measurement reproducibility, where . sequential measurements are advised to be neither monotonically increasing nor decreasing beyond 0.2%.

For IVBT and beta-emitting sources, several measurements will be performed at different source orientations around the chamber cylindrical axis with the obtained results being combined to produce an average value. Since the polarity effect can exceed 0.5% for beta-emitting sources, the applied calibration coefficient is only valid for the well-type chamber polarity stated in the calibration certificate.

8.2.3. Correction for influence quantities

The calibration coefficients for a well-type chamber are only valid under the reference conditions which apply to the calibration. Apart from the source model correction factor, which is treated explicitly in Eq. (12) and Eq. (13), there are a number of other correction factors that need to be applied to the raw reading obtained directly from the electrometer $M_{sm,\ raw}$ to take into account any departure from the reference calibration conditions:

$$M_{sm} = M_{sm,\ raw}\ k_{TP}\ k_{alt}\ k_{leak}\ k_{elec}\ k_{pol}\ k_{s} \tag{27}$$

A brief discussion of each single influence quantity is provided below.

8.2.3.1. Air density correction factor

For high energy photon sources, the standard relation for the air density correction factor k_{TP} given for EBRT can be used:

$$k_{TP} = \frac{(273.15 + T)}{(273.15 + T_0)} \frac{P_0}{P} \tag{28}$$

where P_0 and T_0 are the reference pressure and temperature, respectively, and P and T are the actual pressure and temperature that are recorded at the time of measurement. Application of this relation is only correct if the sensitive air volume of the well-type chamber is vented. It is generally sufficient to check that the vent hole on the side of the chamber is not blocked.

In general, the biggest error in applying k_{TP} comes from using the incorrect reference temperature, since the used P_0 is typically 101.325 kPa. The majority of calibration laboratories use $T_0 = 20°C$, but several countries in North America use $T_0 = 22°C$. Using the wrong reference temperature results in an error of approximately 0.7% for a temperature difference of 2°C.

8.2.3.2. Altitude correction factor

Extra care has to be taken for low energy brachytherapy sources, where Eq. (28) does not fully compensate for significant decreases in ambient pressure taking place at high altitude [148–154]. A modified air density correction factor k'_{TP} is necessary to account for this effect according to:

$$k'_{TP} = k_{TP} k_{\text{alti}} \tag{29}$$

where k_{alti} is an additional altitude correction factor to be included at high altitudes. The magnitude of this correction can be significant, especially for the lowest energy ^{103}Pd photon sources. The effect is device dependent since it depends on the materials and on the design of the air-communicating well-type chambers. On a side note, this phenomenon also applies for low energy X-rays for ionization thimble chambers [155]. Different approaches were described in the literature and some of the obtained results are given below. Users are encouraged to investigate and verify on-site the entity of the correction needed using a calibrated source [88].

Griffin et al. [148] proposed the following altitude correction:

$$k_{\text{alti}} = h_1 P^{h_2} \tag{30}$$

where P is the given pressure [148] and h_1 and h_2 are two parameters to fit the data for the Standard Imaging HDR 1000 Plus or the Standard Imaging IVB 1000 well-type chambers for ^{103}Pd, ^{125}I, and ^{131}Cs low energy sources. The combined relative uncertainty for the altitude correction factor is 0.4% [148]. Values for h_1 and h_2 to be used in Eq. (30) with some seeds, for pressures P in kPa, are given in Table 7.

TABLE 7. PARAMETERS FOR THE ALTITUDE CORRECTION FACTOR k_{ALTI} FOR PRESSURES IN kPa[a] FOR THE STANDARD IMAGING HDR 1000 PLUS AND IVB 1000 WELL-TYPE CHAMBERS [148, 152, 154]

	h_1	h_2
Pd-103 IsoAid Advantage, TheraSeed 200	0.075	0.562
Pd-103 CivaDot	0.073	0.5665
I-125 selectSeed, Theragenics AgX100	0.1095	0.479
I-125 Amersham 6711 (silver rod)	0.1225	0.455
I-125 SourceTech STM 1251 (ceramic)	0.1365	0.431
Cs-131 Caesium Blu (with HDR 1000 Plus)	0.1388	0.4275
Cs-131 Caesium Blu (with IVB 1000)	0.1764	0.3748

[a] Parameters adapted from [148, 152, 154], since (differently to this table) they were originally given for pressures measured in mmHg; the most recent paper by Lambeck [154] gives data in terms of the SI unit (kPa).

Results for the PTW 33005 well-type chamber were provided by Torres del Rio et al. [151] and the following altitude correction factor was proposed for the ^{125}I seeds:

$$k_{alti} = \left[h_3 \left(k_{TP}^{-1} - 1 \right) + 1 \right]^{-1} \tag{31}$$

with $h_3 = -0.476 \pm 0.003$ ($k = 1$). With the same well-type chamber and [103]Pd seeds, an altitude correction was proposed if available for the specific chamber model [153].

The PTW 33004 well-type chamber for HDR sources is not designed for low energy brachytherapy sources. No data on altitude correction factors are therefore available for this chamber.

8.2.3.3. Relative humidity correction

If measurements and calibrations are performed in an adequate *RH* range, no humidity correction factor needs to be applied to the ionization current measured with the well-type chamber. The humidity correction is already applied at the PSDL where air kerma determinations using the primary standard are corrected from normal laboratory conditions (around 50% RH) to reference conditions (dry air, 0% RH).

8.2.3.4. Leakage currents correction factor

Leakage is defined as the signal measured in the absence of a source within the well of the ionization chamber. It can be verified by measuring the signal (current or charge) after having applied the appropriate polarizing voltage for at least 10 minutes. At the SSDL, it is advised to apply a leakage currents correction factor k_{leak} or at least to estimate related uncertainties. Whereas in the hospital, if the leakage signal is below 0.1%, then it can be ignored.

8.2.3.5. Electrometer calibration coefficient

If the electrometer is calibrated separately from the ionization chamber, the electrometer calibration coefficient k_{elec} corrects the electrometer reading to true units of charge/current. The electrometer calibration coefficient is applicable to the range being used on the electrometer. It equals one if the electrometer and ionization chamber are calibrated as a unit, since the electrometer impact is included in the calibration coefficient of the system.

8.2.3.6. Polarity correction factor

The standard practice in PSDLs is to only calibrate well-type chambers at a single polarity. Therefore, no polarity correction is required at the SSDL or clinical setting, as long as the same polarity is applied. Some care is required to ensure that this is the case, and calibration certificates are supposed to be checked to make sure the polarity at calibration is clearly stated. If in doubt, the calibration laboratory can be contacted to get advice on how to set the correct polarizing voltage and the voltage gradient that was used for their well-type chamber. For photon emitting sources, the polarity correction is generally small, but it can be larger for beta-emitting sources and therefore it is important to not ignore the *potential* for a significant polarity effect.

If the electrometer allows polarity selection and the user would like to determine the polarity correction factor k_{pol}, measurements need to be performed with the source dwelling at the sweet spot, using two opposite polarities, for the same integration time (at least 60 s). Before performing new measurements, it is advised to wait at least 10 minutes after each polarizing voltage change. For polarity effect evaluation, the following equation is used:

$$k_{pol} = \frac{|M_+| + |M_-|}{2|M|}$$ (32)

where M_+ and M_- are the electrometer readings obtained at positive and negative polarity, respectively, and M is the electrometer reading obtained with the polarity that is used routinely [40].

8.2.3.7. Ion recombination correction factor

Brachytherapy sources produce a continuous radiation beam. Recombination is usually small for HDR sources (i.e. <0.1%). For ideal conditions (general recombination only), the ion recombination correction factor k_s can be determined with the two-voltage technique [36, 106]:

$$k_s = \left(\frac{4}{3} - \frac{M_1}{3M_2} \right)^{-1}$$ (33)

where M_1 is the electrometer reading at the standard operating voltage for the well-type chamber, V_1, and M_2 is the electrometer reading at $V_2 = V_1/2$. These are generally the only two voltages used, in comparison to k_s measurements for EBRT ionization chambers, and for these settings the IPEM Working Group [141] report states, "For a new ^{192}Ir source with an initial activity of 370 GBq, k_s for the PTW 33004/Nucletron SDS well-type chamber is typically around 1.002, whereas for the Standard Imaging well-type chambers typical values are around 1.001."

Schüller et al. (2015) [145] have reported that Eq. (33) does in principle not apply to PTW Tx33004 well-type chambers of the type $x = W$ or N, where a linear function of $1/I$ versus $1/V$ instead of $1/V^2$ has been observed. The reason for this could be the presence of an undesired collecting volume for this type of chamber which was discovered in a previous study [115]. For the highest accuracy, initial recombination, ion diffusion and the impact of charge screening can be taken into account [156], but these are typically, in total, less than 0.2%.

8.3. SOURCE MODEL CORRECTION FOR AIR KERMA RATE MEASUREMENTS

For HDR and PDR brachytherapy sources, PSDLs and SSDLs are usually limited to the use of specific source models. HDR/PDR brachytherapy sources are handled with remote controlled afterloaders, and calibration laboratories are typically equipped only with one afterloader, which can only be fitted with a specific source model.

According to the formalism given in Section 7, the calibration coefficient includes as a subscript the source model sm_0 used for calibration. This choice points out to the end user the importance of being aware of which type of radiation source was used by the calibration laboratory for calibrating the well-type chamber, compared with the type of source the end user is working with. It is advised to also record this information on the well-type chamber calibration certificate (see Section 6.3).

If the brachytherapy source model used by the end user is different to the model of the source used at the calibration laboratory, the well-type chamber calibration coefficient needs to be multiplied by a source model correction factor, k_{sm,sm_0}, to account for any change of the well-type chamber response because of different source configurations. For the Standard Imaging HDR 1000 Plus well-type chamber with the HDR Iridium Source Holder model 70010, source model correction factors k_{sm,sm_0} have been reported for various HDR and PDR ^{192}Ir source models in two independent studies, a Monte Carlo study performed at the National Physical Laboratory (NPL, UK) [157] and measurements of k_{sm,sm_0} performed at UWADCL [158, 159]. The Monte Carlo method provided a more direct evaluation of the correction factors (with a lower uncertainty than the measured corrections), yet they were supported in magnitude and direction with the measured correction factors. Data were in good agreement within the stated expanded uncertainties ($k = 2$). The Monte Carlo calculated source model correction factors given in Table 8 could form the basis for a future consensus data set. The NPL data set shown in the Shipley et al. 2015 study [157] does not include source model correction factors for the Varian GammaMed Plus PDR ^{192}Ir source. The correction factors for the PDR source were calculated after the publication of the study using the same well-type chamber and source holder model for the Monte Carlo simulation, and also the same formalism as mentioned in [157]. The source model correction factors in Table 8 have an expanded uncertainty of 0.4% ($k = 2$).

Since k_{sm,sm_0} depends on four parameters: (1) the type of source used at the calibration laboratory, (2) the type of source measured by the end user, (3) the type of well-type chamber and (4) the type of source holder, it should be noted that the k_{sm,sm_0} factors listed in Table 8 are only applicable for use with a Standard Imaging HDR 1000 Plus well-type chamber with HDR Iridium Source Holder model 70010. The correction factors are not transferable to different types of well-type chambers and source holders. For other well-type chambers, specific source models and holders, these factors could be calculated based on the formalism in [157]. Monte Carlo techniques validated by experimental measurements are expected to be used.

If the calibration source model is the same as that of the source to be measured, or if no consensus data are available (but sources are constituted by the same radionuclide), the source model correction factor is taken to be unity (i.e. $k_{sm,sm_0} = 1$). In the latter case, an additional uncertainty component has to be added. For currently available sources, the estimation of this additional uncertainty component could be based on the maximum deviation of the source model correction factors for the HDR sources listed in Table 8 (i.e. approximately 2% ($k = 1$)). Strictly speaking, the 2% value is a 'deviation' rather than an 'uncertainty'. However, for this not ideal measurement scenario, the additional uncertainty component in the source model correction factor is advised to be included in the uncertainty budget.

The GammaMed Plus PDR source incorporates an active core of length 0.5 mm, compared to the much longer HDR sources (between 3.5 and 5 mm), which explains the larger deviation from unity of the source model correction factors for the PDR source in Table 8. In this case, the additional uncertainty component for k_{sm,sm_0} might be up to 5% if the well-type chamber was calibrated with one of the HDR sources in Table 8 and then used to measure the source strength of the PDR source or vice versa. As before, the 5% value is based on a 'deviation' rather than an 'uncertainty'.

Data of chamber type-specific radiation quality correction factors, k_Q, were provided by Schüller et al. (2015) to correct for differences in the geometry and the radionuclide between ^{192}Ir and ^{60}Co radioactive sources [145]. The studied well-type chambers were the Standard Imaging HDR 1000 Plus and the PTW 33004/Nucletron SDS chambers (with serial numbers ≥ 315 for PTW 33004 and ≥ 548 for Nucletron SDS). At this stage, this code of practice does not recommend the use of any factors that correct between calibrations of different radionuclides.

Two examples are given to illustrate the use of Table 8.

Example 1: If a calibration laboratory uses an Elekta HDR ^{192}Ir Flexisource (sm_0) to calibrate a Standard Imaging HDR 1000 Plus well-type chamber with a HDR ^{192}Ir source holder model 70010, and the user of the calibrated well-type chamber needs to measure the source strength of a different source model (sm), for instance a Varian VariSource model VS2000, the well-type chamber calibration coefficient shown on the calibration certificate needs to be multiplied by the source model correction factor 0.987 with a standard uncertainty of 0.2% ($k = 1$).

Example 2: On the other hand, if a calibration laboratory uses a Varian VariSource model VS2000 (sm_0) to calibrate a Standard Imaging HDR 1000 Plus well-type chamber with a HDR ^{192}Ir source holder model 70010, and the user of the calibrated well-type chamber needs to measure the source strength of a different source model (sm), for instance an Elekta HDR ^{192}Ir Flexisource, the

TABLE 8. SOURCE MODEL CORRECTION FACTORS, k_{sm,sm_0}, FOR DIFFERENT TYPES OF HDR AND PDR ^{192}IR SOURCES FOR USE WITH A STANDARD IMAGING HDR 1000 PLUS WELL-TYPE CHAMBER WITH HDR IRIDIUM SOURCE HOLDER MODEL 70010, BASED ON MONTE CARLO CALCULATED CORRECTION FACTORS FROM [157] WITH AN EXPANDED UNCERTAINTY OF 0.4% ($k = 2$).

sm \ sm_0	Elekta Flexisource HDR Ir-192	Elekta microSelectron mHDR-v1 (classic)	Elekta microSelectron mHDR-v2	BEBIG HDR Ir-192 GI192 M11	Varian GammaMed Plus HDR	Varian VariSource VS2000	Varian GammaMed Plus PDR[a]
Elekta Flexisource HDR Ir-192	**1.000**	0.996	0.996	0.997	0.999	1.013	1.043
Elekta microSelectron mHDR-v1 (classic)	1.004	**1.000**	1.001	1.001	1.004	1.018	1.047
Elekta microSelectron mHDR-v2	1.004	0.999	**1.000**	1.001	1.003	1.017	1.047
BEBIG HDR Ir-192 GI192M11	1.003	0.999	0.999	**1.000**	1.002	1.016	1.046
Varian GammaMed Plus HDR	1.001	0.996	0.997	0.998	**1.000**	1.014	1.043
Varian VariSource VS2000	0.987	0.983	0.983	0.984	0.986	**1.000**	1.029

TABLE 8. SOURCE MODEL CORRECTION FACTORS, k_{sm,sm_0}, FOR DIFFERENT TYPES OF HDR AND PDR ^{192}IR SOURCES FOR USE WITH A STANDARD IMAGING HDR 1000 PLUS WELL-TYPE CHAMBER WITH HDR IRIDIUM SOURCE HOLDER MODEL 70010, BASED ON MONTE CARLO CALCULATED CORRECTION FACTORS FROM [157] WITH AN EXPANDED UNCERTAINTY OF 0.4% ($k = 2$). (cont.)

sm_0 / sm	Elekta Flexisource HDR Ir-192	Elekta microSelectron mHDR-v1 (classic)	Elekta microSelectron mHDR-v2	BEBIG HDR Ir-192 GI192 M11	Varian GammaMed Plus HDR	Varian VariSource VS2000	Varian GammaMed Plus PDR[a]
Varian GammaMed Plus PDR[a]	0.959	0.955	0.955	0.956	0.958	0.972	**1.000**

[a] Correction factors for Varian GammaMed Plus PDR Ir-192 calculated after publication of [157] by applying the same formalism and using the same geometry of the well-type chamber and source holder for the Monte Carlo simulations.

well-type chamber calibration coefficient shown on the calibration certificate needs to be multiplied by the source model correction factor 1.013 with a standard uncertainty of 0.2% ($k = 1$).

8.4. SHORT TERM REPEATIBILITY CHECKS OF THE WELL-TYPE CHAMBER INSTRUMENTATION

Monitoring of the stabilization behaviour of the well-type chamber dosimetry system can also form part of the regular QA procedures, as any change in this behaviour would indicate a potential operational problem. The short term stabilization behaviour can be validated by acquiring data immediately when a source is inserted inside the chamber and continuing until a stable response is obtained. Temperature of the well-type chamber and the air pressure have also to be monitored continuously and k_{TP} applied to the raw reading.

To establish baseline data for parameters such as short term repeatability, a larger number of measurements are carried out. The relative standard deviation of these readings with respect to the mean reading is verified and eventually investigated if outside the defined limits. It is also useful to vary the acquisition time of measurements to investigate the linearity of the electrometer.

8.5. LONG TERM STABILITY CHECKS OF THE WELL-TYPE CHAMBER INSTRUMENTATION

A check of the well-type chamber stability with time ensures that the system is operating properly and that the measurements are compatible with those made at the time of calibration. It is advised to perform stability checks for the well-type chamber instrumentation on a regular basis both at the SSDL and the hospital, at least four times per year, and before and after each source exchange.

In principle, the recommended method to provide the highest level of confidence in a brachytherapy dosimetry system is based on complete redundancy of equipment (i.e. backup well-type chamber and source holder, electrometer, extension cable, thermometer/barometer/hygrometer), with the two systems being regularly compared (Section 8.5.1). In practice, it is not always achievable to have access to a redundant independent well-type chamber dosimetry system, particularly at hospitals. Therefore, another recommended method to check chamber long term stability is to use mechanically stable *check sources* (Section 8.5.2).

In the absence of a redundant well-type chamber dosimetry system or of an adequate check source, chamber response can be monitored with the two other alternative methods provided in Sections 8.5.3 and 8.5.4 (not recommended at SSDLs). In general, for those methods that compare chamber response to a baseline value (i.e. check source, external radiation beam and HDR source), constancy of corrected readings is advised to be within 0.5% from the baseline. The baseline should be defined at the time of the well-type chamber commissioning and at each recalibration. It is advised to investigate any discrepancy greater than 0.5% [16, 113, 118], with well-type chamber recalibration as a possible option.

8.5.1. Redundant well-type chamber dosimetry system method

When the same radiation source is measured with two independent systems and all measurements are decay corrected to the same reference time, the ratios of the corrected ionization currents from different well-type chambers is advised to remain constant within typically ±0.1% of the running mean. If the change in the numerical value of the ratio is outside this range this may indicate a problem with one or both well-type chambers. Investigation is required, for instance with one of the other methods presented in this section (preferably check sources), to address the problem.

Redundancy generally refers to two independent systems maintained by a single institution and compared so that performance can be monitored. An option that might be available to clinical users, but not recommended for SSDLs, is to compare systems from two different institutions, or locations, using the same radiation source. The assumption here is that agreement implies that both systems are operational, which is reasonable but not as rigorous a check as either a redundant system at the same facility or a check source. However, such a comparison also facilitates discussions of equipment maintenance and usage, measurement procedures, data analysis, etc. and can potentially identify improvements and/or potential failure modes so is valuable. Inter-centre comparisons may certainly be considered as part of the wider QA practices of any institution.

8.5.2. Check source method

A *check source* provides a very reliable check on system operation. Check sources are recommended to be mechanically stable, with possibly a long half-life and an energy comparable to that of the analysed source. For high energy sources (i.e. ^{192}Ir, ^{137}Cs and ^{60}Co), ^{137}Cs is the optimal radionuclide, due to its half-life and energy, for producing small tubes to be used as check sources. Analogously, to check the chamber constancy for low energy LDR sources (e.g. ^{103}Pd, ^{125}I and ^{131}Cs), ^{241}Am is suggested [36]. Due to the lower ionization current produced in the well-type chamber with ^{241}Am, a higher statistical uncertainty compared to measurements with a ^{137}Cs source is expected. Since it is not always easy to obtain [241]Am in a geometry suitable for insertion into a well-type chamber, ^{137}Cs check sources can be used to check chamber stability for low energy sources.

The check source will be inserted into the well-type chamber using a dedicated holder/spacer to perform the constancy verification. It is important to have a reproducible set-up, with the check source placed on the central axis of the well-type chamber. The reference distance from any reference point of the well-type chamber (e.g. entrance or bottom of the chamber) is supposed to be fixed, with the check source position being close to the sweet spot for chamber response. Rotation of the holder/spacer with respect to the well-type chamber needs to also be kept constant, unless they are symmetrical to the long central axis (e.g. with response constancy shown to be within 0.05%). Corrections are advised to be applied for temperature and pressure as well as for the decay of the check source.

8.5.3. External radiation beam method

For this method, a 6 MV linac or a ^{60}Co teletherapy unit can be used [16, 118, 160]. In this case, the radiation field is very different in both energy and dimensions, and therefore the measurement is not as directly correlated with the measurement of brachytherapy sources. The chamber is placed on the ground at a fixed extended-SSD distance. The chosen field will expose the entire well-type chamber and minimally include the triaxial cable (to reduce extracameral current), with the electrometer positioned far from the primary radiation field of the EBRT source (i.e. >3 m). The treatment couch should be completely retracted to not interfere with the primary beam, and no trays or other beam attenuating devices should be present. The key is to have a reproducible set-up geometry. Irradiation times are supposed to be of the order of a few minutes, enough to minimize the effect of beam-on effects and/or shutter timer errors, but short enough to avoid over-ranging the electrometer. The beam output should be known in accordance with the local dosimetry protocol. Since there is no direct correlation between the measurement of external radiation beam output and the source strength determination of a brachytherapy source, this method serves only to define a baseline value that can be used for monitoring long term chamber stability. Reference temperature, pressure and relative humidity need to be recorded at the time of baseline reading, and it is advised to correct the well-type chamber readings for these environmental conditions each time a verification is performed. Averaging over several measurements can eventually replace initial baseline reading.

8.5.4. HDR source method

This method can be used only in the absence of a redundant well-type chamber dosimetry system (see 8.5.1) and of check sources (see 8.5.2). It can be used for chamber stability monitoring over a period of a few months with ^{192}Ir sources [161] and over a period of a few years for ^{60}Co sources. The HDR source installed in the afterloader can in principle be used as a check source to demonstrate system stability until the old source is replaced. The same procedure as used for source calibration is followed, and the decay corrected result is compared with the original measurement at installation. This technique relies on good repeatability of positioning of the source within the chamber, and accurate knowledge of the radionuclide half-life. Such a method provides at least confidence in the ability to correctly characterize the new source to be installed.

8.6. SOURCE EXCHANGE AND THE VENDOR SOURCE CERTIFICATE

8.6.1. Ordering and exchanging a source

Brachytherapy sources are ordered using a process that clearly documents and verifies the accuracy of the source strength, number of sources, radionuclide, shipment address(es) and regulatory aspects. For receipt of brachytherapy source(s) it is customary to perform radiation surveys, wipe tests, and to confirm delivery with a national authority. Survey meters with valid calibrations for the particular radiation quality should be used to determine the highest dose rate (mSv/h) in contact with the package and at 1 m from the surface for comparison to a transport index value (if labelled on the package). The correct labelling of the package, such as the shipper and addressee, label type, radionuclide listed, and activity (units of Bq), is expected to be checked.

Depending on the local regulations, either the radiation safety officers, medical physicists, metrologists or other responsible professionals in hospitals and SSDLs will test the used equipment that was in contact with the source for the presence of radionuclide contamination. A wipe test of the package exterior and interior will be used to check for possible contamination. For photon emitting sources, the minimum detectable activity (MDA) is 185 Bq (5 nCi). The physicist is supposed to have access to a wipe test system that is calibrated with MDA and system settings specific to the ordered radionuclide. Using modern NaI well type scintillators, photomultipliers, and scalers employing lower-level discriminators to enhance signal-to-noise, background and wipe test counts can be performed efficiently while satisfying the MDA requirements. The last task is to update the local inventory with the activity received, applicable date and number of sources. Generally, the manufacturer-stated activity is used for inventory purposes. Other specific tasks may be required in accordance with the national laws. The disposal of the old source(s) is also expected to follow adequate recommendations and regulations.

8.6.2. The role of the vendor source certificate at the hospital

Brachytherapy sources are accompanied with a source certificate provided by the vendor stating the source strength, as determined by the manufacturer. The traceability to a primary standard has to be stated. It is advised to use the contained source activity only for licensing, inventory, and transportation purposes, and is not relevant to the clinical source strength determination or dose calculation [88, 142]. Source strength measurements performed by a third-party (separate from the manufacturer or medical physicist) are discouraged as a means of satisfying the requirement for independently measured source strength [25].

8.6.2.1. HDR source calibration

The source strength is advised to be measured by the clinically qualified medical physicist and then used as the reference input for the afterloader treatment console (and the TPS) [162]. It is important to perform an independent measurement using a traceably calibrated well-type chamber dosimetry system, and according to a national or international code of practice [25, 37, 118, 119, 123, 141, 163]. In particular, it is recommended that the source strength of each single HDR photon-emitting brachytherapy source is measured, prior to its clinical use. The procedure described in the previous paragraphs are expected to be followed according to the current code of practice.

For HDR sources, the typical uncertainty of the source strength stated in the manufacturer's certificate is 5% ($k = 3$), providing a coverage probability of approximately 99.7%. For a normal distribution, this is equivalent to an uncertainty of 3.3% ($k = 2$), providing a coverage probability of approximately 95%. Based on the uncertainties of well-type chamber calibration coefficients, which can be achieved with current calibration methods, the discrepancy between the source strength stated in the source calibration certificate and that measured by the medical physicist is typically less than 3%. Discrepancies >3% are advised to be investigated. If discrepancies >5% are observed, it is recommended to not use the source clinically, until the differences have been reconciled [123, 141]. Source strength values on the source certificate and those measured by the medical physicist have to agree within their stated expanded uncertainties with a coverage factor $k = 2$.

If a physical quantity different from the measured one is required by the afterloader treatment control console and/or the TPS, the measured quantity will be appropriately converted to the required one.

8.6.2.2. LDR sources

For LDR sources such as low-energy photon-emitting seeds, the clinical workflow does not always permit source strength measurements preceding seed implantation [88]. It may not be possible to measure the seeds due to them being shipped in sterile cartridges or strands. Therefore, it is strongly recommended that additional seeds, obtained from the same batch as those to be implanted, are ordered and calibrated [25]. These additional seeds will be ordered at the same time as the seeds to be implanted and could be shipped to a different location to facilitate the measurement.

For permanent LDR implants, there are generally two prescriptions, pre-implant (to facilitate the seed order) and post-implant. Practically, the ordered source strength is included in the pre-implant prescription, which may slightly differ from the values included in the manufacturer's calibration certificate and that measured by the medical physicist. These seeds are manufactured in batches, and the average source strength is reported by the manufacturer. The clinically qualified medical physicist is advised to independently compare the nominal and measured source strengths of the source batch using a traceably calibrated well-type chamber dosimetry system. If the difference between the mean measured source strength for a sources assay of at least five seeds and the value given in the manufacturer's source certificate is within 5%, the sources can in principle be used for clinical purposes. If the difference is higher than 5%, it is advised to extend the sources assay by a further five seeds and the comparison repeated. If the source sample cannot be extended or the discrepancy is confirmed after increasing the number of measured sources of the same batch, it is advised to discuss this discrepancy with the manufacturer [88]. The radiation oncologist should be consulted to decide about the clinical use of this source batch.

The post-implant prescription includes the radionuclide, number of seeds, and the total source strength implanted. The source strength that is ordered is generally derived from a nomogram to estimate the implanted conditions. The source strength measured by the medical physicist is used in the post-implant prescription, and preferably in the pre-implant treatment plan and during the intraoperative treatment planning process. In analogy to HDR sources, if a physical quantity different from the RAKR is required by the TPS, an appropriate RAKR conversion to this quantity is expected to be done.

For temporary LDR implants such as with ^{125}I seeds or ^{137}Cs tubes, the medical physicist measures the source strength, and it is expected that this source strength value, with appropriate decay correction, will be used for all patients. In the case of individual sources, a difference between the measured source strength and the one stated by the manufacturer up to 6.0% is acceptable [25, 88]. For differences higher than 6%, it is important that the radiation oncologist is consulted to decide about the use of this source. Some other circumstances not considered in this section are outlined in other publications [25].

8.6.2.3. *Beta emitting sources*

Beta-emitting brachytherapy sources are generally used for ophthalmic applicators, eye plaques or IVBT sources. The number of laboratories offering calibrations for these sources has greatly decreased in the past decades because of their decreased clinical usage. Yet there are new sources recently introduced to the marketplace for use in eye plaques. These include concave ^{106}Ru/^{106}Rh eye plaques of varying dimensions with calibration standards under development. Once there is an established system of calibrations, a parallel plate ionization chamber and reference geometric set-up will permit calibrations in the clinic. The proposed approach is outlined in Hansen, et al. [53].

Well-type chambers for ^{90}Sr/^{90}Y IVBT sources are still being calibrated. The 30 mm source train is being calibrated, giving the absorbed dose to water calibration coefficient with its appropriate uncertainty. A well-type chamber with an appropriate sweet spot length is supposed to be used for the calibration.

9. ESTIMATED UNCERTAINTIES IN THE DETERMINATION OF THE REFERENCE AIR KERMA RATE UNDER REFERENCE CONDITIONS

Since the mid-1990s, many PSDLs have developed air kerma primary standards for LDR brachytherapy sources (^{103}Pd, ^{125}I, ^{131}Cs and ^{192}Ir) and HDR brachytherapy sources (^{60}Co and ^{192}Ir) [77] (see Section 6). Depending on the measurement method, source type and primary standard used, the relative standard uncertainties ($k = 1$) in the measurement of the RAKR (or AKS) of brachytherapy sources estimated by different PSDLs, range from 0.8% to 1.3% for LDR sources and from 0.6% to 1.5% for HDR sources. For SSDLs, typical standard uncertainties ($k = 1$) range from 0.9% to 1.5% for LDR sources and from 0.7% to 1.7% for HDR sources.

The uncertainties that affect the different physical quantities or procedures contributing to the overall RAKR determination can be divided in several steps over the entire standard dissemination chain. The combination in quadrature of the uncertainties resulting from the different steps yields the combined standard uncertainty. Examples of estimates of the uncertainty levels achievable in the RAKR measurement are provided in Table 9 for an LDR ^{125}I source, based upon content from [82, 142, 164–166], and in Table 10 for an HDR ^{192}Ir source. These tables list relative standard uncertainties of physical quantities or procedures that are used for the whole traceability chain from the measurement of LDR ^{125}I or HDR ^{192}Ir brachytherapy sources by the end users back to the PSDL level. Two illustrative scenarios for the establishment of uncertainty budgets are in both cases presented. As recommended by this code of practice, well-type chambers are used as reference instruments.

The physical quantities and procedures given in Table 9 and Table 10, as well as values for the percentages listed, are only given to provide an example. It is essential that the end users perform their own uncertainty evaluation based on their own measurement procedures and equipment. It is not possible to present a generic uncertainty analysis that can be used by all users. As a standard approach, it is recommended that the uncertainty values taken into consideration are supported by their evidence. This is preferably achieved by deriving these values from quality control tests of the measuring instrumentation. Alternatively, those specified by the manufacturer can be used. In practice, for a specific task, it is often recommended to first define the desired uncertainty level and then take appropriate measures to obtain it.

More information about the formalism is provided in the Evaluation of Measurement Data - Guide to the Expression of Uncertainty in Measurement [167] and in Appendix VI. Furthermore, IAEA-TECDOC-1585, Measurement Uncertainty [168] provides guidance to SSDLs on assessing and reporting measurement uncertainties related to their calibration services.

For Table 9 that deals with an LDR ^{125}I brachytherapy source, scenario 1 describes a case where a reference class dosimeter is used, and its performance complies or exceeds the requirements of this code of practice. It describes the case where the irradiation conditions are tightly controlled (i.e. in terms of seed positioning, air density, choice of correction factors, etc.) and the relevant corrections for influence quantities are applied.

Scenario 2 of Table 9 describes a *possible but sub optimal situation* where the measurement conditions are not controlled or monitored, and some corrections are not implemented. For example, the chamber has not reached thermal equilibrium, air pressure is not accurately measured, and the signal is noisy without proper background subtraction. In both scenarios it is assumed that correction factors are properly applied for a specific source model. Assuming standard uncertainties for the dose rate constant Λ of 2.5% [82] and 3.5% for scenarios 1 and 2, respectively, starting from the $\dot{K}_{\delta,R}$ values given in Table 9 the expanded uncertainty ($k = 2$) associated with the determination of the absorbed dose to water to the reference point, $\dot{D}_W(r_0,\theta_0)$, would result in 5.4% and 8.0%, respectively.

For Table 10 that deals with an HDR ^{192}Ir brachytherapy source, scenario 1 describes a case where a reference class dosimeter is used, and its performance complies or exceeds the requirements of this code of practice. It describes the case where the irradiation conditions are tightly controlled (i.e. in terms of source model, source positioning, air density, etc.) and the relevant corrections for influence quantities are applied. In this scenario, the same source model is used in calibrations and hospital measurements and therefore no source model corrections or related uncertainty are needed.

Scenario 2 of Table 10 describes a *realistic but sub optimal situation* where the measurement conditions are not controlled or monitored, and some corrections are not implemented. For example, the source used at the SSDL is different from the PSDL and a source model correction factor from Table 8 is used. A standard uncertainty related to the use of this correction factor is included in the uncertainty budget (0.2% in Table 10, step 2, scenario 2). If no published source model correction factors are available, for instance if the hospital uses a different type of well-type chamber or a new type of ^{192}Ir source which is not listed in Table 8, an additional uncertainty component will be added for scenario 2 (see Table 10, step 4, scenario 2). It is not possible to account for any future source or well-type chamber designs, but based on the largest source model correction factor for the existing HDR ^{192}Ir sources listed in Table 8 (i.e. 1.018 for the VariSource VS2000/Elekta microSelectron-v1 combination), which is equivalent to a 1.8% deviation, an uncertainty component of at least 2% is advised to be considered (see Table 10, step 4, scenario 2). Strictly speaking, the 2% value (rounded up from 1.8%) is a 'deviation' rather than an 'uncertainty'. However, for this not ideal measurement scenario the additional uncertainty component in the source model correction factor in Table 10, step 4, scenario 2 needs to be included. The uncertainty component might be even higher, for instance for PDR sources where the source geometries can be quite different compared to HDR sources (as mentioned in Section 8.3).

Assuming standard uncertainties for the dose rate constant Λ of 0.5% [82] and 5% for scenarios 1 and 2, respectively, starting from the $\dot{K}_{\delta,R}$ values given in Table 10 the expanded uncertainty ($k = 2$) associated to the determination of the absorbed dose to water to the reference point, $\dot{D}_W(r_0,\theta_0)$, would result in 1.9% and 11.5%, respectively.

TABLE 9. EXAMPLE OF THE ESTIMATE OF THE RELATIVE STANDARD UNCERTAINTY OF $\dot{K}_{\delta,R}$ MEASURED WITH A CALIBRATED WELL-TYPE CHAMBER, FOR A TYPICAL LDR ^{125}I SEED[a]

Physical quantity or procedure[b]	Relative standard uncertainty (%)	
	Scenario 1	Scenario 2
Step 1: RAKR calibration of reference standard at PSDL (in RAKR)		
Establishment of the calibration coefficient	0.8	1.3
Combined uncertainty (step 1)	**0.8**	**1.3**
Step 2: RAKR measurement at SSDL with reference standard [see Eq. (12)]		
Long term stability of secondary standard	0.1	0.2
Current measurement	0.2	0.4
Set-up and seed positioning	0.2	0.3
Ion recombination correction factor	0.1	0.2
Temperature and pressure correction factor	0.1	0.5
Impact of humidity	0.1	0.3
Combined uncertainty (steps 1 + 2)	**0.9**	**1.5**
Step 3: definition of calibration coefficient of the well-type chamber dosimetry system to be calibrated [see Eq. (15)]		
Current measurement	0.2	0.4
Set-up and seed positioning	0.2	0.3
Ion recombination correction factor	0.1	0.2
Radioactive decay correction factor	0.1	0.2
Temperature and pressure correction factor	0.1	0.5
Impact of humidity	0.1	0.3
Combined uncertainty (steps 1 + 2 + 3)	**0.9**	**1.7**
Step 4: RAKR measurement at hospital with calibrated well-type chamber dosimetry system [see Eq. (12)]		
Long term stability of the calibrated chamber	0.1	0.2
Current measurement	0.2	0.4

TABLE 9. EXAMPLE OF THE ESTIMATE OF THE RELATIVE STANDARD UNCERTAINTY OF $\dot{K}_{\delta,R}$ MEASURED WITH A CALIBRATED WELL-TYPE CHAMBER, FOR A TYPICAL LDR ^{125}I SEED[a] (cont.)

Physical quantity or procedure[b]	Relative standard uncertainty (%)	
	Scenario 1	Scenario 2
Set-up and seed positioning	0.2	0.3
Ion recombination correction factor	0.1	0.2
Temperature and pressure correction factor	0.2	0.5
Impact of humidity	0.1	0.3
Combined standard uncertainty for $\dot{K}_{\delta,R}$	**1.0**	**1.9**
Expanded uncertainty ($k = 2$) for $\dot{K}_{\delta,R}$	**2.0**	**3.8**

[a] This table provides an example of uncertainty estimation for an LDR ^{125}I seed. It is essential that the end users perform their own uncertainty evaluation based on their own measurement procedures and equipment.

[b] A relative standard uncertainty of 0.4% needs to be included for any of the four steps in the uncertainty budget where measurements are performed at high altitudes or very low pressures, where an altitude correction factor needs to be applied. The additional uncertainty components and all the other uncertainty components listed in Table 9 will then have to be added in quadrature to calculate a revised combined standard uncertainty.

TABLE 10. EXAMPLE OF THE ESTIMATE OF THE RELATIVE STANDARD UNCERTAINTY OF $\dot{K}_{\delta,R}$ MEASURED WITH A CALIBRATED WELL-TYPE CHAMBER, FOR A TYPICAL HDR [192]IR SOURCE[a]

Physical quantity or procedure	Relative standard uncertainty (%)	
	Scenario 1	Scenario 2
Step 1: RAKR calibration of reference standard at PSDL (in RAKR)		
Establishment of the calibration coefficient	0.6	1.5
Combined uncertainty (step 1)	**0.6**	**1.5**

TABLE 10. EXAMPLE OF THE ESTIMATE OF THE RELATIVE STANDARD UNCERTAINTY OF $\dot{K}_{\delta,R}$ MEASURED WITH A CALIBRATED WELL-TYPE CHAMBER, FOR A TYPICAL HDR [192]IR SOURCE[a] (cont.)

Physical quantity or procedure	Relative standard uncertainty (%)	
	Scenario 1	Scenario 2
Step 2: RAKR measurement at SSDL with reference standard [see Eq. (12)]		
Long term stability of secondary standard	0.1	0.2
Current measurement	0.1	0.3
Set-up and source positioning	0.2	0.3
Ion recombination correction factor	0.1	0.2
Temperature and pressure correction factor	0.1	0.5
Impact of humidity	0.1	0.3
Source model correction factor	0	0.2
Combined uncertainty (steps 1 + 2)	**0.7**	**1.7**
Step 3: definition of calibration coefficient of the well-type chamber dosimetry system to be calibrated [see Eq. (15)]		
Current measurement	0.1	0.3
Set-up and source positioning	0.2	0.3
Ion recombination correction factor	0.1	0.2
Radioactive decay correction factor	0.1	0.2
Temperature and pressure correction factor	0.1	0.5
Impact of humidity	0.1	0.3
Combined uncertainty (steps 1 + 2 + 3)	**0.7**	**1.9**
Step 4: RAKR measurement at hospital with calibrated well-type chamber dosimetry system [see Eq. (12)]		
Long term stability of the calibrated chamber	0.1	0.2
Current measurement	0.1	0.3
Set-up and source positioning	0.2	0.3
Ion recombination correction factor	0.1	0.2

TABLE 10. EXAMPLE OF THE ESTIMATE OF THE RELATIVE STANDARD UNCERTAINTY OF $\dot{K}_{\delta,R}$ MEASURED WITH A CALIBRATED WELL-TYPE CHAMBER, FOR A TYPICAL HDR ^{192}IR SOURCE[a] (cont.)

Physical quantity or procedure	Relative standard uncertainty (%)	
	Scenario 1	Scenario 2
Temperature and pressure correction factor	0.2	0.5
Impact of humidity	0.2	0.3
Source model correction factor	0	2
Combined standard uncertainty for $\dot{K}_{\delta,R}$	**0.8**	**2.8**
Expanded uncertainty ($k = 2$) for $\dot{K}_{\delta,R}$	**1.7**	**5.7**

[a]   This table provides an example of uncertainty estimation for an HDR ^{192}Ir source. It is essential that the end users perform their own uncertainty evaluation based on their own measurement procedures and equipment.

10. APPLICATION OF REFERENCE QUANTITIES IN THE HOSPITAL

Brachytherapy dose calculations are based in general upon a consistent formalism that utilizes dosimetry parameters for uniform dose delivery across the globe. Components of the formalism usually include a measure of source strength for determining the output for a specific brachytherapy source, as well as dosimetry parameters that apply to a given source model. This approach assumes constant manufacturing practices such that dosimetric characterization of a particular source model at any one point in time will apply to all sources of the same model.

To avoid dose calculation errors, it is strongly recommended to use only the quantities and units endorsed in Section 3 of this Code of Practice for the specification of brachytherapy sources. As previously pointed out in IAEA-TECDOC-1274, Calibration of Photon and Beta Ray Sources Used in Brachytherapy: Guidelines on Standardized Procedures at Secondary Standards Dosimetry Laboratories (SSDLs) and Hospitals [36], extra care is needed when converting quantities which have already been converted. For example, if the source calibration carried out by the manufacturer is directly traceable to a standards laboratory, but the source strength on the source certificate is shown with a different quantity, the quantity on the certificate has first to be converted back to the calibration quantity by dividing by the conversion coefficient which was used by the manufacturer. Only then, the conversion to the desired quantity can be performed in one step. If this procedure is not followed, there is a risk that the manufacturer has based the conversion of the source strength quoted on the source certificate on a different factor than that used by the end user and that the traceability of the source strength is lost.

Application of the calibration quantities for the different brachytherapy sources is described below, including an introduction of approaches for dose distribution calculations. Part of the provided information is summarized in Table 11.

10.1. PHOTON-EMITTING RADIOACTIVE SOURCES

The dosimetry formalism used worldwide in brachytherapy TPSs, for most of the intracavitary, interstitial and intraluminal applications delivered with photon-emitting HDR and LDR sources, is based upon the AAPM TG-43 report [23, 79–83]. Key to this report is its use of the calibrated source strength. The dose rate at any given location in water is directly proportional to the source strength and other influencing quantities as illustrated below in Eqs (34) and (35).

$$\dot{D}(r,\theta) = S_K \Lambda \frac{G_X(r,\theta)}{G_X(r_0,\theta_0)} g_X(r) F(r,\theta) \tag{34}$$

$$\dot{D}(r,\theta) = \dot{K}_{\delta,R} \Lambda_{r_0} \frac{G_X(r,\theta)}{G_X(r_0,\theta_0)} g_X(r) F(r,\theta) \tag{35}$$

where Eq. (35) represents a variation to the original TG-43 formalism, and $\dot{K}_{\delta,R}$ instead of S_K is used to express the source strength. These equations are applicable for a 2D dosimetric characterization of a brachytherapy source, with the reference coordinate system provided in Fig. 1.

In both cases, given the dose rate constants Λ or Λ_{r_0}, the dose rate in water at the reference point P(r0, θ0) is obtained according to Eqs (3) and (4), respectively. The other terms in Eqs (34) and (35) are influencing quantities of the dose rate:

(a) Geometry function $G_X(r, \theta)$
(b) Radial dose function $g_X(r)$

TABLE 11. RECOMMENDED CALIBRATION QUANTITIES FOR THE DIFFERENT BRACHYTHERAPY SOURCES AND RELATED DOSE DISTRIBUTION CALCULATION MODALITIES

Radiation type	BT clinical applications	Recommended calibration quantities	Possible dose distribution calculation methods
Photon (radionuclide)	Intracavitary, interstitial, intraluminal, ophthalmic	$\dot{K}_{\delta,R}, S_K,$ $\dot{D}_{W,R}$	TG-43, adapted TG-43, or Monte Carlo
	Surface		Hand calculation, library plan, TG-43, Monte Carlo, based on an ionization chamber measurement [181]
Beta	Intravascular	$\dot{D}_W(2\ \text{mm})$	PDD-based
	Ophthalmic	—[a,b]	PDD-based, adapted TG-43, or Monte Carlo
Photon (eBT)	Intracavitary	—[a,c]	Adapted TG-43 or Monte Carlo
	Surface		PDD based or Monte Carlo

[a] Not provided in this code of practice.
[b] Useful information can be found in Appendix IV.
[c] Useful information can be found in Appendix III.

(c) 2D Anisotropy function $F(r, \theta)$

where X is substituted with P or L to indicate if the point or line source approximation is chosen, respectively. In some cases, a simplified version of Eq. (34) and of Eq. (35) with the 1D anisotropy function $\Phi_{an}(r)$ instead of $F(r, \theta)$ is chosen. A more extensive description of the TG-43 formalism and of each one of these functions is given in Appendix V. The German standard organization Deutsches Institut Für Normung (DIN) published Norm DIN-6803-2, Dosimetry for Photon Brachytherapy – Part 2 [169], which introduces a specific calibration coefficient for the well-type chamber that incorporates Λ aligning to the TG-43 formalism in all other aspects.

Dose distribution calculation in surface brachytherapy delivered with cone-shaped applicators (e.g. Valencia and Leipzig applicators) [170–177] is usually based on hand calculation or on pre-calculated 2D dose distribution data that have to be rescaled by the actual source strength [29, 178]. Library data for Nucletron HDR ^{192}Ir sources and Valencia and Leipzig applicators can be found on-line [179]. Dosimetric characterization of the source dwelling in the applicator, with the plastic cap placed over the end of the applicator, can be achieved with small-volume parallel plate ionization chambers [180, 181].

Flap applicators (e.g. Freiburg applicator), which are generally used to treat wider lesions than those treated with cup-shaped applicators, usually are employed in combination with the TG-43 algorithm [29].

Alternate approaches for calculating the dose starting from the TG-43 formalism can be proposed for intracavitary eBT applications [182], with the source strength being either measured directly in terms of $\dot{D}_{W,R}$ [183], or $\dot{D}_{W}(r_0, \theta_0)$ being obtained by applying a specific factor to the physical quantity characterizing the source strength [184].

10.2. BETA-EMITTING RADIOACTIVE SOURCES

Given their different radiological properties than photon emitting sources, dosimetry for beta-emitting brachytherapy sources is not easily characterized using the TG-43 formalism because of the non-exponential dose fall-off. Characterization is complicated further for planar sources such as eye plaques or for spinal dural treatments where the radionuclide distribution is not accurately estimated by a point or line segment [185, 186]. As such, dose calculations are generally limited to point or 1D depth-dose determinations [28].

For line segment sources such as beta-emitting IVBT, the AAPM TG-60 and TG-149 reports provide a recommended approach for normalizing dose at a depth of 2 mm and performing dosimetry using a Cartesian coordinate system instead of the polar coordinate system inherent to the TG-43 dose calculation formalism [26, 187].

For eye plaques, an approach modifying the TG-43 dose calculation formalism has been developed for clinical treatment planning [27, 188].

While source calibration and dose calculation methods for beta-emitting sources lag in comparison to those of photon emitting sources, it is worth noting their potential advantages for conformal treatments.

10.3. BRACHYTHERAPY SOURCE REGISTRIES

Good practice and a quality management system is required for brachytherapy services, with brachytherapy sources needing robust calibration and dosimetry standards. Global harmonization requires confirmation that a manufacturer source calibration program is traceable to a primary calibration standard through transference of the calibration standard to an SSDL (or ADCL), source strength comparisons to demonstrate the accuracy and constancy of the calibration program have been conducted, and the source dosimetry parameters (e.g. dose rate constant) have been evaluated using two independent methods [189, 190].

For a source meeting these dosimetry prerequisites, the source manufacturer (or a clinical user) may apply for consideration of the source on the Brachytherapy Source Registry (BSR). The BSR is jointly managed by the AAPM and the Imaging and Radiation Oncology Core (IROC) Houston QA Center. Review of an application for posting on the BSR is performed by the AAPM BSR Working Group, who then makes a recommendation to the AAPM Brachytherapy Subcommittee. The sources included in the BSR, the AAPM dosimetry prerequisites and other important material are available on-line [89]. Manufacturers who do not comply with the dosimetry prerequisites are requested to maintain compliance but removed from the BSR if insufficient actions are taken.

The European Society for Radiotherapy and Oncology (ESTRO) maintains a database of brachytherapy sources that includes the dosimetry parameters in a convenient spreadsheet format [91], and sources that have not yet or no longer meet the AAPM dosimetry prerequisites. Harmonization of these two registries is maintained through joint participation and leadership of the BSR Working Group by both AAPM and ESTRO members.

10.4. TYPICAL UNCERTAINTIES IN PATIENT DOSIMETRY

Uncertainties in the dose delivery process might influence the clinical outcome, in terms of both local control and side effects. In both LDR and HDR brachytherapy treatments, overall clinical uncertainties are estimated as the combination of individual uncertainty contributions, which are related to several parameters, such as:

(a) Source strength calibrations traceable to a PSDL;

(b) Λ, G_X, g_X, F or Φ_{an} data estimations used for treatment planning calculations and interpolations, if the TG-43 algorithm is applied. Other possible parameter estimations if dose calculation methods different than the TG-43 algorithm are used;

(c)   Imaging techniques and applicator/catheter/source position placement and reconstruction;

(d)   Target contouring (including intra- and inter-observer variability);

(e)   Tissue heterogeneity and patient finite dimensions;

(f)   Applicator absorption;

(g)   Dose delivery;

(h)   Anatomy variations with time.

Uncertainties depend on the clinical application and are in general different with different radioactive sources (i.e. LDR or HDR, low energy or high energy, radionuclide, model, calibration laboratory), treated anatomical regions, possible fractionations, and level of adaptation according to image-guidance. Exhaustive discussions, tables and examples are provided in DeWerd et al. [142] and Kirisits et al. [19]. It is recommended that each institution performs adequate comprehensive uncertainty estimations by identifying, quantifying and grouping together uncertainty components that affect each specific clinical brachytherapy treatment process. Sub optimal elements at any stage of the treatment preparation and delivery process can be thus identified and possibly improved in terms of diminishing their dosimetric uncertainty.

Appendix I

ANTIQUATED QUANTITIES AND UNITS

Brachytherapy has worked with a number of antiquated quantities that users have insisted on continuing using. Use of these units can cause errors because of conversion factors. These quantities may result in up to 10% errors. One of the quantities that is persistently used is apparent activity A_{app}. According to its definition, A_{app} is derived from the RAKR, that is traceable to the appropriate standard, according to the following equation:

$$A_{app} = \frac{\dot{K}_{\delta,R} d_R^2}{(\Gamma_\delta)_K} \tag{36}$$

where d_R is the reference distance and $(\Gamma_\delta)_K$ is the air kerma rate constant[5]. A_{app} depends on the RAKR; it cannot be experimentally determined independently. The value of $(\Gamma_\delta)_K$ is needed, which depends on the source model (i.e. radionuclide, construction of the source and its encapsulation). As already stated about 20 years ago in IAEA-TECDOC-1274 [36], since "different air kerma rate constants have been published for many brachytherapy sources, failure to uniformly define and apply $(\Gamma_\delta)_K$ could cause significant confusion and unnecessary treatment delivery errors". This quantity is therefore not expected to be used for dosimetry purposes.

Some governments require declaration of contained activity for transportation purposes. This number is not accurate and is therefore not expected to be used for any clinical application. Measurement of the contained activity is tenuously correlated to the A_{app} through corrections for the source encapsulation. The same values of A_{app} for two different sources containing the same radionuclide do not necessarily correspond to the same values of contained activity since their relationship depends on the source design. Also, becquerel is the SI unit for activity, not curie and or the equivalent mass of radium, mgRaEq.

In the past, jigs for free air measurements have been used for calibrating HDR ^{192}Ir sources [16, 36, 161]. This technique is no longer to be used. Uniformity and accuracy of measurements are superior when using a well-type air ionization chamber.

---

[5] The index δ in the air kerma rate constant $(\Gamma_\delta)_K$ indicates that only photons with energies greater than δ are taken into account. Photons with energies below this threshold are considered to be absorbed in the radionuclide (i.e. auto-absorption) or in its encapsulation.

ESTABLISHMENT OF PRIMARY CALIBRATION STANDARDS
FOR RADIOACTIVE BRACHYTHERAPY SOURCES

The present situation for dosimetry standards based on air kerma and absorbed dose to water for the sources considered in this code of practice is summed up in Table 12 and Table 13 at the end of this section, respectively. Sections II.1 and II.2 give a brief overview of different brachytherapy dosimetry standards, either primary standards or instruments directly traceable to primary standards, which are either currently in use in different national metrology institutes (NMIs) around the world or which have been developed as prototypes. A more detailed description of these instruments can be found in two recent review articles [77, 144] and specific articles and reports mentioned in the list of references.

II.1. PHOTON-EMITTING RADIOACTIVE SOURCES

Since the early 1990s, many NMIs have developed air kerma primary standards for different types of photon-emitting brachytherapy sources, such as ^{103}Pd, ^{125}I, ^{192}Ir, ^{131}Cs, ^{137}Cs and ^{60}Co. Usually, $\dot{K}_{\delta,R}$ or S_K of the high energy sources are realized with ionometric standards based on ionization chambers, since they show a relatively large signal-to-noise ratio. $\dot{D}_W$ can either be realized with an ionometric standard and a conversion to absorbed dose rate to water via a Monte Carlo calculated conversion factor, or by a more direct measurement based on absorbed dose calorimetry or chemical dosimetry.

II.1.1. Air kerma dosimetry standards for LDR sources

The National Institute of Standards and Technology (NIST, USA) has established a cylindrical wide-angle free air chamber (WAFAC) for the realization of the quantity air kerma strength for LDR ^{103}Pd, ^{125}I and ^{131}Cs sources [164, 165]. The LDR sources are set up at 30 cm distance from the 8 cm diameter aperture of the WAFAC. A 0.1 mm thick aluminium filter between the source and the aperture removes the low energy fluorescence X-rays, which originate in the titanium encapsulation of the sources. According to the AAPM TG-43U1 report [81] and ICRU report 72 [87], only photons with energies greater than delta are considered for the definition of S_K and $\dot{K}_{\delta,R}$, respectively, because photons with energies less than δ contribute only insignificantly to the absorbed dose at depths >1 mm in tissue. The value of δ is typically 5 keV for LDR sources and 10 keV for HDR sources [23, 81]. Air kerma strength measurements with the WAFAC can be performed for photon energies up to 40 keV. This involves the measurement of ionization currents from two different collecting volumes and the application of various conversion and correction factors.

Calibrations of LDR ^{192}Ir and ^{137}Cs sources are performed at NIST using spherical, graphite-walled cavity ionization chambers [191].

UWADCL has developed a variable-aperture free air chamber (VAFAC) for measuring S_K of LDR ^{103}Pd, ^{125}I and ^{131}Cs sources [166]. The VAFAC has a large diameter collecting electrode, can be operated in an extrapolation mode and is used for brachytherapy sources with photon energies up to 70 keV. The chamber is similar in design to NIST's WAFAC. The variable aperture of the VAFAC, however, also allows the study of the angular dependence of air kerma strength measurements.

The National Research Council (NRC, Canada) also commissioned a primary standard WAFAC based on the NIST design to measure S_K of LDR ^{103}Pd and ^{125}I seeds. For both the NRC WAFAC and the UW VAFAC, a considerable polarity effect was observed. This effect was eliminated by covering both the front and rear surfaces of the collecting electrodes of the NRC WAFAC and UW VAFAC with an electrically conducting material [192]. However, it was also shown that this did not have a direct impact because the differential measurement of the charge collected from two collecting volumes removes the polarity effect.

The Physikalisch-Technische Bundesanstalt (PTB, Germany) has established a large-volume parallel-plate extrapolation chamber (GROVEX) as $\dot{K}_{\delta,R}$ primary standard for LDR ^{103}Pd and ^{125}I photon sources [193]. The design of the GROVEX is similar to the NIST WAFAC. However, the separation between the two parallel-plate electrodes is adjusted automatically. For several plate separations, which are larger than the range of the secondary electrons, the ionization currents are measured, and the air kerma rate is obtained from the gradient of this function, when all Monte Carlo calculated correction factors have been applied. The GROVEX extrapolation chamber is named for traditional reasons, based on a series of research projects in Germany and at PTB [194]. However, it has never been intended to use the GROVEX as an extrapolation chamber in the meaning that the measurements are extrapolated to zero plate separation, under Bragg-Gray conditions. In fact, the measurements are performed within secondary electron equilibrium at sufficiently large plate separations.

The Laboratoire National Henri Becquerel (LNE-LNHB, France) at the Commissariat à l'Energie Atomique et aux Energies Alternatives (CEA) has developed a circular free air chamber in the shape of a torus with an outer radius of around 50 cm as a primary standard for LDR ^{103}Pd and ^{125}I sources [195]. The rectangular cross section of the torus has the features of a conventional free air chamber. For the $\dot{K}_{\delta,R}$ measurement, the LDR sources are placed inside a Kapton tube either with or without a 0.1 mm thick tubular aluminium filter at the centre of the circular chamber. The advantage of the circular design of the chamber is that the $\dot{K}_{\delta,R}$ measurements are non-sensitive to source positioning. Any effects due to source anisotropy will be averaged during the measurement.

NPL uses a spherical three-litre NE2551 protection level ionization chamber to measure $\dot{K}_{\delta,R}$ of LDR ^{125}I seeds and LDR ^{192}Ir wires [196, 197]. The ionization chamber is traceably calibrated against NPL's primary standard free air chambers for low and medium energy X rays and the ^{60}Co therapy level primary standard cavity chamber. The three-litre ionization chamber is calibrated in 25 keV and 33 keV X ray beams from the ISO 4037-1 (1993) narrow spectrum series [198] to derive the ionization chamber's ^{125}I calibration coefficient by calculating the average of the 25 keV and 33 keV calibration coefficients. The ionization chamber is also calibrated in 35 keV to 250 keV X rays and ^{137}Cs and ^{60}Co gamma rays. The chamber's ^{192}Ir calibration coefficient is obtained by weighting the chamber's different energy responses according to the ^{192}Ir spectrum.

The $\dot{K}_{\delta,R}$ standard of the Istituto Nazionale di Metrologia delle Radiazioni Ionizzanti of Ente per le Nuove tecnologie l'Energia e l'Ambiente (ENEA-INMRI, Italy) is based on an interpolation technique, developed by Verhaegen et al. [199], where three spherical ionization chambers are traceably calibrated against ENEA's air kerma primary standards in X rays from the ISO 4037-1 series and ^{60}Co gamma rays [200]. The $\dot{K}_{\delta,R}$ of LDR ^{125}I and ^{192}Ir, and also of HDR ^{192}Ir sources, is measured with these ionization chambers.

At the D. I. Mendeleev All-Russian Institute for Metrology (VNIIM, Russian Federation), S_K measurements of LDR ^{125}I seeds are performed using both a PTW type TM32002 one-litre spherical ionization chamber and an ATOMTEX type BDKR-01M scintillation detector. The instruments are traceably calibrated against VNIIM's air kerma primary standard for X rays in the energy range from 16 keV to 33 keV at ISO 4037 N20–N40 and L20–L35 radiation qualities. The ^{125}I calibration coefficients are determined by linear interpolation [201, 202].

II.1.2. Air kerma dosimetry standards for HDR sources

No primary standard for HDR ^{192}Ir sources had been established until the beginning of the 1990s. In 1991, Goetsch et al. [203] developed an interpolation method for the calibration of ionization chambers at the UW-ADCL to measure the S_K of HDR ^{192}Ir sources. The average photon energy of a typical HDR ^{192}Ir source is approximately halfway between the effective energy of the NIST M250 X-ray beam quality (146 keV) and ^{137}Cs gamma rays (662 keV). A therapy level cavity ionization chamber from the UW ADCL was traceably calibrated in air in these two reference beams against the NIST primary standards and Goetsch et al. obtained the chamber's ^{192}Ir calibration coefficient by linear interpolation. The calibrated cavity chamber was used to measure S_K of an HDR ^{192}Ir source at several distances in air. Measurements were taken at seven source-to-chamber distances between 10 cm and 40 cm to determine correction factors for positioning errors and room scatter. A thorough review of the seven-distance technique and a description of an improved version of the original seven-distance apparatus is given by Rasmussen et al. and Stump et al. [159, 204].

The seven-distance technique was recommended by the IAEA [36]. Many PSDLs and SSDLs developed similar methods for $\dot{K}_{\delta,R}$ or S_K measurements of HDR ^{192}Ir sources during the 1990s and early 2000s. In 2006, Mainegra-Hing and Rogers [205] from the NRC improved the accuracy of the seven-distance technique by interpolating the calibration coefficient for ^{192}Ir based on $1/N_K$ values, not N_K values. Additionally, it was shown that the wall correction factors in the Goetsch interpolation method would not be needed. The NRC primary standard for HDR ^{192}Ir brachytherapy sources [137] is based on a spherical graphite ionization chamber where the ^{192}Ir calibration coefficient was determined using the approach recommended by Mainegra-Hing and Rogers [205] in which the arithmetic mean was taken of the inverse of the calibration coefficients for a 250 kV narrow spectrum X ray beam (N250) and a ^{137}Cs gamma ray beam, directly traceable to the NRC primary standards for these two radiation beams. This approach was used for a number of afterloaders by Rasmussen et al [159].

Further, the simplified analytical methods have been devised to evaluate the scatter contribution and distance error required for $\dot{K}_{\delta,R}$ or S_K determination of HDR ^{192}Ir brachytherapy sources using a seven-distance technique and a Farmer-type cylindrical ionization chamber by Kumar et al. [206, 207].

A refined version of the interpolation method for the UWADCL air kerma strength standard, a further modification of the seven-distance apparatus and an uncertainty budget were presented by Rasmussen et al. [159].

At VNIIM, S_K measurements of HDR ^{192}Ir sources are performed using a PTW type TM32005 30 cm^3 spherical ionization chamber, which is traceably calibrated against VNIIM's air kerma primary standards for X rays and gamma radiation using the CCRI 250 radiation quality (effective energy 124 keV) and ^{137}Cs, respectively. The chamber's ^{192}Ir calibration coefficient is determined by linear interpolation [202].

Rather than just relying on two photon beam qualities, other national measurement institutes calibrate their ionization chambers for a range of additional medium energy X rays and also for ^{60}Co gamma rays before interpolating to [192]Ir. Further interpolation methods for the calibration of cavity ionization chambers for HDR ^{192}Ir have been established at PTB [136, 208, 209], ENEA-INMRI [200], and the Radiological Science Laboratory of Rio de Janeiro State University (LCR, Brazil) [210, 211]. Interpolation methods for the calibration of NE2571 thimble type chambers for HDR ^{192}Ir have been implemented at LNE-LNHB [212, 213], the Van Swinden Laboratorium (VSL, The Netherlands) [134, 214, 215] and the Australian Radiation Protection and Nuclear Safety Agency (ARPANSA, Australia) [216, 217]. At LNE-LNHB, an NE2571 thimble chamber is rotated around an HDR ^{192}Ir source at different radii, typically ranging from 100 mm to 220 mm source-to-detector distance, with a high accuracy of ±52 µm.

The NPL has established a cavity ionization chamber as primary standard for HDR ^{192}Ir [135, 218]. The volume of the air cavity was measured as part of the commissioning. The NPL method to measure $\dot{K}_{\delta,R}$ is based on the Bragg-Gray principle and the application of large cavity theory, and it does not require an interpolation of calibration coefficients because the cavity chamber has been directly commissioned for the gamma spectrum of a commercially available HDR ^{192}Ir brachytherapy source.

Bhabha Atomic Research Centre (BARC, India) has also established a graphite cavity ionization chamber as primary standard based on a known collecting volume for standardization of HDR ^{192}Ir brachytherapy sources in terms of $\dot{K}_{\delta,R}$. The BARC method for measuring $\dot{K}_{\delta,R}$ is based on the Burlin general cavity theory. The air kerma calibration coefficient (N_K) of this ionization chamber was estimated analytically using cavity theory and also validated with Monte Carlo calculations [219]. Other primary standard graphite cavity chambers based on known collecting volumes and the application of cavity theory for the measurement of $\dot{K}_{\delta,R}$ or S_K of HDR ^{192}Ir sources have been developed at the Institute of Nuclear Energy Research (INER, Taiwan) [220], the Korea Research Institute of Standards and Science (KRISS, Republic of Korea) [221] and the National Metrology Institute of Japan (AIST-NMIJ) [138]. The National Institute of Metrology (NIM, China) has also developed a primary standard graphite cavity chamber with known collecting volume, which is currently being commissioned for HDR ^{192}Ir.

II.1.3. Absorbed dose to water dosimetry standards for LDR sources

An alternate approach to calibrating conventional brachytherapy sources instead of using S_K or $\dot{K}_{\delta,R}$ is to directly measure the reference absorbed dose rate to water at 1 cm from the source along the source transverse plane $\dot{D}_W(r_0,\theta_0)$. The quantities, r_0 and θ_0, are specified according to the coordinate system provided in Fig. 1. A direct measurement of $\dot{D}_W(r_0,\theta_0)$ eliminates the need to convert air kerma to dose by applying the dose rate constant Λ. This more direct approach potentially reduces the overall uncertainty on the absorbed dose to water.

Each of three PSDLs (ENEA-INMRI, LNE-LNHB and PTB) developed ionometric absorbed dose to water primary standards for LDR ^{125}I brachytherapy photon sources as part of the joint research project T2.J06, 'Increasing cancer treatment efficacy using three-dimensional (3D) brachytherapy', which is within the framework of the European Metrology Research Programme (EMRP) from 2008 to 2011. Absorbed dose calorimetry at room temperature was not considered for LDR sources because of the expected low measurement signal. All three absorbed dose standards for LDR ^{125}I seeds were therefore based on ionometry because large signal to noise ratios could be achieved.

The PTB has designed and built an in-phantom free air chamber (ipFAC) for LDR ^{125}I and ^{103}Pd sources. This large-volume parallel-plate extrapolation chamber was previously known as 'GROVEX II' [222]. However, in this case the measurement is not based on the extrapolation method, hence a new name was found to reflect the actual measurement method. The entrance plate and the back plate of the chamber are made of water-equivalent material (RW1). During the measurement, the LDR sources are inserted into a small RW1 cylinder and rotated around their long axis at 30 cm distance from the entrance plate. A Monte Carlo calculated conversion factor is applied to the difference of the ionization charges collected at two different plate separations to yield $\dot{D}_W(r_0, \theta_0)$ [223, 224].

At ENEA-INMRI a large-angle variable-volume ionization chamber (LAVV-1) was developed to be used as an LDR absorbed dose rate standard [225]. The chamber is embedded in a high-purity graphite phantom and operates under 'wall-less air chamber' conditions. The measurement method is similar to the one developed at PTB for the ipFAC [223].

For absorbed dose measurements of LDR ^{125}I sources, LNE-LNHB uses the same circular free air chamber which was described in Section II.1.1 [195], but with a modified source holder. The LDR seeds are placed either inside a water equivalent PMMA sphere with 1 cm radius or a hollow Kapton cylinder with 1 cm radius, filled with liquid water. Each of the source holders can be set up at the centre of the circular ionization chamber. The measured water kerma rate is finally converted to $\dot{D}_W(r_0, \theta_0)$ by applying a Monte Carlo calculated conversion factor.

At UWADCL a cryogenic calorimeter with a liquid helium thermal sink was developed. They managed to measure the emitted power from low energy, low dose rate brachytherapy sources [226, 227].

II.1.4. Absorbed dose to water dosimetry standards for HDR sources

Absorbed dose calorimeters measure the heating effect of ionizing radiation in a medium, (e.g. water or graphite). For HDR brachytherapy sources, as opposed to LDR sources, the dose rate is large enough so that calorimeters can be established as absorbed dose primary standards. The feasibility of water calorimetry for HDR ^{192}Ir sources was demonstrated at McGill University (Montréal, Canada) [76, 78].

Absorbed dose calorimeters for HDR ^{60}Co and/or ^{192}Ir sources were also developed within the EMRP T2.J06 project. PTB and VSL modified their existing absorbed dose water calorimeters for external beam radiotherapy. The main design of PTB's water calorimeter has previously been described in the literature [228]. To enable absorbed dose measurements of HDR brachytherapy sources, a new source holder with a stainless steel needle to hold the sources at a distance of 2.5 cm from the calorimetric measurement point with an uncertainty below 100 μm was built [229].

The VSL's HDR ^{192}Ir absorbed dose standard is a modified version of the water calorimeter for external ^{60}Co and MV photon beams [230]. VSL's newly developed source holder for the water calorimeter contains an aluminium heat sink for dealing with the source self-heating effect of the radioactive HDR ^{192}Ir source.

Both ENEA-INMRI and NPL independently developed and built two graphite calorimeters for HDR ^{192}Ir as part of the EMRP T2.J06 project. Both calorimeters contain a ring-shaped graphite core with 2.5 cm mean radius which is surrounded by a vacuum gap. The core in ENEA's calorimeter is surrounded by two annular graphite jackets with further vacuum gaps between the components to limit heat transfer from the core to the environment [231]. NPL's HDR ^{192}Ir calorimeter contains two graphite tubes that are positioned between the source and the core, separated by further vacuum gaps to limit any conductive heat transfer between the source and the core, and also from the core to the environment [232].

The absorbed dose standards described in this section are currently not used to calibrate secondary standard instruments [77]. However, at the end of the EMRP T2.J06 project the $\dot{D}_{W}(r_0,\theta_0)$ standards were used together with the $\dot{K}_{\delta,R}$ standards from the four PSDL's mentioned in Section II.1.1 and II.1.2 to measure the dose rate constants of different types of LDR ^{125}I and HDR ^{192}Ir brachytherapy sources and good agreement within the stated uncertainties was found with published consensus values [233, 234].

A different approach for the realization of $\dot{D}_{W}(r_0,\theta_0)$ for HDR ^{192}Ir is via chemical dosimetry using a Fricke system [235]. The NRC has developed a ring-shaped ferrous sulphate Fricke device for the absolute measurement of $\dot{D}_{W}(r_0,\theta_0)$ for HDR ^{192}Ir sources [236]. The Fricke system is less sensitive to temperature changes compared to calorimetry. Further information on Fricke systems for brachytherapy applications can be found in the scientific literature [236–240].

Some of the air kerma rate and absorbed dose rate measurement methods described in Sections II.1.1 to II.1.3 take into account the source anisotropy by using a ring-shaped detector. Otherwise, either the source or the detector is rotated during the measurement. In some cases, measurements are taken from either one or two directions, and S_{K}, $\dot{K}_{\delta,R}$ or $\dot{D}_{W}(r_0,\theta_0)$ are determined from the average of multiple source transfers.

II.2. BETA-EMITTING RADIOACTIVE SOURCES

Calibrations of brachytherapy beta sources where a well-type chamber cannot be used are not available at an SSDL, except for the UWADCL [53, 241]. ^{90}Sr/^{90}Y and ^{106}Ru/^{106}Rh ophthalmic applicators can also be calibrated at some PSDLs, (e.g. NIST and NPL). Guidance on the calibration of brachytherapy beta sources can be found in ISO 21439 [242]. The ISO standard is applicable to sealed radioactive sources, for example, planar and curved ophthalmic applicators, where only the beta radiation emitted is relevant for delivering a dose.

The performed calibration is traceable to the NIST primary standard with a relative expanded uncertainty of 20% ($k = 2$) and is performed in a water phantom by a plastic scintillator detector with 1 mm diameter and 0.5 mm height [188].

NPL operates a calibration service for curved and planar ^{90}Sr/^{90}Y and ^{106}Ru/^{106}Rh ophthalmic applicators traceable to NPL's ^{60}Co absorbed dose to water primary standard graphite calorimeter via alanine dosimetry. Calibration of the ophthalmic applicators is in terms of dose rate at 0 mm and 2 mm depth in water along the plaque central axis with a relative expanded uncertainty of 7% ($k = 2$). The calibration is performed by irradiating cylindrical alanine pellets of 0.5 mm thickness and 5 mm diameter, which are placed on a PMMA phantom. A depth-dose curve is measured with a stack of ten alanine pellets and the surface dose rate at 0 mm is determined by an extrapolation of the curve to 0 mm thickness. The UWADCL measured the curved ^{106}Ru/^{106}Rh eye plaques with a windowless extrapolation chamber [53].

High dose rate ^{90}Sr IVBT sources have been calibrated traditionally by clinical end users from source strength measurements using a well-type chamber. Globally, there no longer is a primary calibration standard available for ^{90}Sr IVBT sources. However, medical physicists may still obtain well-type chamber calibrations for IVBT from the UWADCL [243], which maintains a calibration standard that is traceable to NIST and demonstrates system constancy since closure of the NIST primary calibration standard. Clinical users of ^{90}Sr sources for pterygium (non-cancerous growth on the cornea of the eye) are unable to obtain calibrations as most PSDLs no longer maintain primary calibration standards for ^{90}Sr ophthalmic applicators (except NPL) and no secondary laboratories offer calibrations. However, clinical users may demonstrate constancy (when corrected for decay) of their source output through measurements in a well-type chamber and comparisons of output from other long-lived sources.

Table 12 and Table 13 summarize all dosimetry standards discussed in Appendix II.1 and Appendix II.2.

TABLE 12. CURRENT STATUS OF BRACHYTHERAPY AIR KERMA DOSIMETRY STANDARDS[a]

Radionuclide (dose rate)	Reference standard	Reference standard methodology	Laboratory	Work in progress
Ir-192 (HDR/PDR)	Cavity ionization chamber	Interpolation and multiple distance method	ARPANSA, ENEA-INMRI, LCR, LNE-LNHB, NRC, PTB, UWADCL, VNIIM, VSL	
	Cavity ionization chamber with measured air cavity volume	Bragg-Gray principle and calculated correction factors, multiple distance method	AIST-NMIJ, BARC, INER, KRISS, NIM, NPL	NIM's new HDR ^{192}Ir primary standard cavity chamber currently being commissioned
Co-60 (HDR)	Cavity ionization chamber	Multiple distance method	LNE-LNHB, PTB	LNE-LNHB setting up HDR ^{60}Co calibration service
I-125 seeds (LDR)	Cavity ionization chamber	Interpolation method	ENEA-INMRI, NPL	

TABLE 12. CURRENT STATUS OF BRACHYTHERAPY AIR KERMA DOSIMETRY STANDARDS[a] (cont.)

Radionuclide (dose rate)	Reference standard	Reference standard methodology	Laboratory	Work in progress
	Wide angle free air chamber (WAFAC)	Two-volume measurement technique	NIST, NRC	
	Circular free air chamber	Free-in-air charge measurement	LNE-LNHB	
	Cavity ionization chamber and scintillation detector	Interpolation method	VNIIM	
	Large-volume extrapolation chamber (GROVEX)	Multiple-volume measurement technique	PTB	
	Variable aperture free air chamber (VAFAC)	Multiple-volume measurement technique	UWADCL	
Pd-103 seeds (LDR)	WAFAC, VAFAC, GROVEX	See descriptions for LDR [125]I	NIST, NRC, PTB, UWADCL	
Cs-131 seeds (LDR)	WAFAC, VAFAC	See descriptions for LDR [125]I	NIST, UWADCL	
Cs-137 tube sources (LDR)	Cavity ionization chamber with measured air cavity volume	Bragg-Gray principle and calculated correction factors	NIST	

97

TABLE 12. CURRENT STATUS OF BRACHYTHERAPY AIR KERMA DOSIMETRY STANDARDS[a] (cont.)

Radionuclide (dose rate)	Reference standard	Reference standard methodology	Laboratory	Work in progress
Ir-192 wires (LDR)	Cavity ionization chamber	Interpolation method	ENEA-INMRI, NPL	
	Cavity ionization chamber with measured air cavity volume	Bragg-Gray principle and calculated correction factors	NIST	

[a] The instruments listed are either primary or secondary standards and are of the highest metrological quality. Strictly speaking, the ionization chambers where the chamber factor is derived using an interpolation technique (via traceable calibrations against primary standards), are secondary standards.

TABLE 13. CURRENT STATUS OF BRACHYTHERAPY ABSORBED DOSE TO WATER DOSIMETRY STANDARDS

Radionuclide (dose rate)	Reference standard	Reference standard methodology	Laboratory	Work in progress
Ir-192 (HDR/PDR)	Absorbed dose water calorimeter	Measurement of ionizing radiation heating	PTB, VSL	
	Absorbed dose graphite calorimeter	Measurement of ionizing radiation heating	ENEA-INMRI, NPL	
	Ring-shaped Fricke device	Spectrophotometric measurement of change in optical density	NRC, LCR	
Co-60 (HDR)	Absorbed dose water calorimeter	Measurement of ionizing radiation heating	PTB	
I-125 seeds (LDR)	Large-angle variable-volume ionization chamber (LAVV-1)	Two-volume measurement technique	ENEA-INMRI	
	Circular free air chamber with water-equivalent source holder	Water kerma rate measurement, conversion to D_w	LNE-LNHB	

TABLE 13. CURRENT STATUS OF BRACHYTHERAPY ABSORBED DOSE TO WATER DOSIMETRY STANDARDS (cont.)

Radionuclide (dose rate)	Reference standard	Reference standard methodology	Laboratory	Work in progress
	In-phantom free air chamber (ipFAC)	Multiple-volume measurement technique	PTB, VNIIM	New absorbed dose primary standard for LDR ^{125}I based on PTB's and ENEA's designs currently being developed at VNIIM
	Liquid helium calorimeter [226]	Electrical substitution technique, measurement of emitted power	UWADCL	Calorimeter being modified to reduce uncertainty in measured emitted power of LDR sources
Pd-103 seeds (LDR)	In-phantom free air chamber (ipFAC)	Multiple-volume measurement technique	PTB	
Sr-90/Y-90 (LDR) for IVBT	Indirect absorbed dose to water standard	Extrapolation chamber	NIST	Use decreased substantially
Sr-90/Y-90 and Rh-106/ Ru-106 (LDR) for eye plaques	Plastic scintillation detector	Scintillation measurement	NIST	

TABLE 13. CURRENT STATUS OF BRACHYTHERAPY ABSORBED DOSE TO WATER DOSIMETRY STANDARDS (cont.)

Radionuclide (dose rate)	Reference standard	Reference standard methodology	Laboratory	Work in progress
	Graphite calorimeter	Alanine pellets calibrated against graphite calorimeter, conversion to D_w	NPL	
Rh-106/Ru-106 (LDR) for eye plaques	Convex windowless extrapolation chamber	Multiple-volume measurement technique	UWADCL	Extrapolation chamber at the UWADCL

Appendix III

X-RAY EMITTING ELECTRONIC SOURCES

III.1. INTRODUCTION

In addition to radionuclide-based photon-emitting radiation sources, miniature low energy X rays emitting electronic sources are emerging as a possible radiotherapeutic treatment delivery system. Since no radioactive sources are involved, efforts for radiation protection, transport and safety are reduced [24]. Regular operating rooms can often be used without any further shielding due to the short photon range. At the time of drafting this publication there are a few known companies marketing miniature X ray tubes as medical devices. A brief description of these systems is given below.

The INTRABEAM® PRS500 system (Carl Zeiss Surgical, Oberkochen, Germany) is a compact X ray source originally developed for intracranial treatments [244, 245]. The electrons are accelerated by an accelerator unit and directed through a drift tube by a control unit. On the inside of the probe tip, the electrons hit a gold target, producing X rays [246–248]. Some of the electrons are scattered back and detected by an internal monitor. This information is used twice, first to adjust the beam position to achieve an isotropic distribution and secondly to precisely monitor the X ray yield during the irradiation time. Accessories are supplied for dosimetry and quality assurance [249]. This includes a water phantom to validate the depth dose curve supplied by the vendor. For the dose measurements a PTW soft X-ray chamber type 34013 is used. For intraoperative radiotherapy (IORT) of breast cancer, spherical plastic applicators were developed to irradiate the tumour bed. Other applicators have been developed for gynaecological, skin and kyphoplasty treatments [250–252].

The Xoft® Axxent® system (iCAD Inc., Nashua, NH) is a miniature X ray tube integrated into a multi-lumen catheter along with a cooling sheath, first introduced in 2006. Measured and simulated dosimetry parameters for this device were described by Rivard et al. [253] and Liu et al. [254]. The strength of the source is checked before each treatment with a well-type chamber adapted for radiation protection purposes. Early breast cancer has been treated using balloon catheters, first with multiple fractions [255] and then with a single fraction IORT [256]. Dosimetric properties were also described for endometrial [257], surface [258] and intracavitary applicators [259]. Power variations between sources, flatness and symmetry for surface applicators were all within 5%. The most homogeneous radiation is perpendicular to the emitter axis, which is why the dose reference point is also defined there. There is an air kerma rate traceable standard developed at NIST for this source (see Section III.2).

Ariane Medical Systems Ltd (Derby, UK) has already introduced two devices into the market: Papillon 50 and Papillon +.

Papillon 50 was launched in 2008 [24] and is primarily used for contact radiation therapy. Electrons are accelerated in an evacuated copper tube and hit a rhenium transmission target. In the further course of the tube there is an ionization chamber for dose monitoring, an Al-flattening filter and an exit window made of polycarbonate. The spatial distribution of the applied radiation has a fixed aperture angle of 45°. Thus, the dose reference point is also located along the emitter axis. The measured and Monte Carlo simulated relative dose distributions for this device were described by Croce et al. [260]. The system is primarily used as a boost for EBRT of rectal cancer. However, the system also offers applicators for skin irradiation [24].

The Papillon + system [261] was developed with the aim of also being suitable for intracavitary applications (i.e. to achieve the most isotropic radiation distribution possible and at the same time a high dose rate). A two-shell construction was used. Inside, there is a tube with a diameter of 10 mm. At the end of the tube is a rounded beryllium cap, which is coated with tungsten on the inside. The beryllium cap is electrically separated from the rest of the tube by a ceramic insulator. The electrons hitting the tungsten target are discharged again via an anode current measuring device, whereby the beam current is also recorded during treatment. Around this inner tube is a cooling cover. Between the two tubes is a cooling circuit based on mineral oil. For dosimetry application specific phantoms designed for a PTW TM23342W ionization chamber are provided for the end user to facilitate measurement of absorbed dose according to AAPM guidelines. The dose reference points are aligned along the source axis.

Wolf-Medizintechnik presented the ioRT-50 device which can be used for intercavitary, skin and contact therapy. It is based on a hollow anode tube with a maximum high voltage of 70 kV and a maximum tube current of 7 mA. The entire tube body is water-cooled, and the built-in radiation shielding is made of lead-free materials. The focal spot has a diameter of 15 mm and the angular distribution of the emittance field of the bare tube is $180° \times 360°$. The dose reference point is also located along the axis of this tube.

The Esteya® EB system (Elekta AB-Nucletron, Stockholm, Sweden) is operated at about 69.5 kV and was developed for the treatment of skin lesions [262]. A QA device is used to check the consistency of power, flatness, and depth dose. Surface applicators with a flattening filter are used to achieve a dose distribution similar to the one obtained with the Valencia ^{192}Ir HDR applicators from the same manufacturer. Dosimetry of the device has been described by Garcia-Martinez et al. [263].

Photoelectric therapy (Xstrahl Ltd, Camberley, United Kingdom) was launched in late 2014 and is also aimed at treating skin lesions. The system consists of a mobile equipment with built-in cooling, collimation device and flattening filters for a consistent dose profile. Dedicated QA tools are available to allow ionization chambers placement for dose rate verification [24].

III.2. PRIMARY STANDARDS FOR X-RAY EMITTING SOURCES

In general, across different manufacturers, there is discord and no uniform calibration standard for the devices listed above. For most of them, calibration techniques currently rely on external-beam photon calibration standards [264]. However, these approaches are less robust and consequently provide larger uncertainties when determining the absorbed dose to the patient. Well-type chambers are in many cases not suitable, because the diameter of the well at the centre of the chamber housing might not be wide enough. Other types of ionization chambers, for instance thin-window parallel-plate ionization chambers with suitable holders which can be traceably calibrated against primary standards, might be more appropriate for performing measurements close to these devices, except that the use of a parallel-plate chamber does not account for the 360-degree aspect of the source.

At the time of drafting this code of practice, NIST is the only national metrology institute that developed a primary calibration standard and that is offering a calibration service for at least one single source type of the Xoft Axxent System. The primary standard is optimized for a reference point perpendicular to the source axis which is suitable for the emittance field characteristic of the source. The suggested quantity by NIST is the reference air kerma rate from the source transverse plane at 50 cm in air, $\dot{K}_{air,50}$, "so as to avoid the relatively large added uncertainty associated in this case with air-kerma strength. This choice can change depending on possible future developments in clinical-dosimetry protocols by the AAPM" [265]. $\dot{K}_{air,50}$ is measured with NIST's Lamperti free air chamber [266] and has units Gy s^{-1}. This calibration standard may be transferred to a well-type chamber having a holder specific to this model of eBT source. Well-type chambers for the Xoft Axxent system, traceable to NIST, have source holders that are specifically designed for them. Information about the current status of air kerma dosimetry standards for X-ray emitting sources is given in Table 14.

Absorbed dose to water primary standards for some of the electronic brachytherapy sources listed in the previous section are being developed as part of the European Metrology Programme for Innovation and Research (EMPIR) project 'Primary standards and traceable measurement methods for X-ray

TABLE 14. CURRENT STATUS OF ELECTRONIC BRACHYTHERAPY AIR KERMA DOSIMETRY STANDARDS[a]

eBT source (dose rate)	Reference standard	Reference standard methodology	Laboratory	Work in progress
Xoft Axxent (HDR)	Lamperti free air chamber	Free-in-air charge measurement	NIST	New source model being worked on by NIST
	Attix free air chamber	Free-in-air charge measurement	UWADCL	
INTRABEAM, Xoft Axxent (HDR)	Free air chamber	Free-in-air charge measurement	VSL	Being developed/ modified as part of EMPIR project PRISM-eBT

[a] The instruments listed are either primary or secondary standards and are of the highest metrological quality. Strictly speaking, the ionization chambers where the chamber factor is derived using an interpolation technique (via traceable calibrations against primary standards), are secondary standards.

emitting electronic brachytherapy devices' (PRISM-eBT) [183]. The dose reference point will be at 1 cm distance along the axis given by the emittance characteristic of the specific source type. As part of the project, LNE-LNHB has developed a methodology for standardization of electronic brachytherapy sources in the unit of absorbed dose to water [267]. Table 15 provides more information about the current status of absorbed dose to water dosimetry standards for X-ray emitting sources. To determine the absorbed dose to water at a reference depth of 1 cm in a water phantom, a conversion factor is calculated using the Monte-Carlo method. The method was exemplified by the calibration of a 4 cm spherical applicator of the Zeiss INTRABEAM system. The authors determined an absorbed dose value which was significantly higher than the value specified by the manufacturer. According to the authors this finding is consistent with similar observations published in the literature [268–270].

TABLE 15. CURRENT STATUS OF ELECTRONIC BRACHYTHERAPY ABSORBED DOSE TO WATER DOSIMETRY STANDARDS

eBT source (dose rate)	Reference standard	Reference standard methodology	Laboratory	Work in progress
INTRABEAM, Xoft Axxent, (HDR)	Free-air chamber	Free-in-air charge measurement, conversion to D_w	CMI[a]	Being developed/ modified as part of EMPIR project PRISM-eBT
	Free-air chamber	Free-in-air charge measurement, conversion to D_w	ENEA-INMRI	Being developed/ modified as part of EMPIR project PRISM-eBT
	Circular free air chamber with eBT source holder	Free-in-air charge measurement, conversion to D_w	LNE-LNHB	Being developed/ modified as part of EMPIR project PRISM-eBT
	In-phantom free air chamber (ipFAC): Parallel-plate ionization chamber in a plastic phantom	Multiple-volume measurement technique; source at 30 cm distance	PTB	Being developed as part of EMPIR project PRISM-eBT
	In-water ion chamber (IWIC): Parallel-plate ionization chamber combined with a water phantom	Multiple-volume measurement technique; source at 1 cm distance in the water phantom	PTB	Being developed as part of EMPIR project PRISM-eBT

[a] The Cesky Metrologicky Institut (CMI) is the Czech Metrology Institute.

Appendix IV

OTHER DETECTOR SYSTEMS FOR BRACHYTHERAPY

Although the well-type chamber is the recommended method for performing brachytherapy measurements for most available sources at SSDLs and hospitals, there are sources for which that is not the case. The measurement based on well-type chambers do not apply to the measurements/calibrations of sources used with surface applicators or beta particle planar and concave sources used in ophthalmic brachytherapy. Some of the detectors other than the well-type chamber that might be considered are radiochromic films, thermoluminescence dosimeters (TLDs), diodes, diamond detectors, alanine detectors, optically stimulated luminescence dosimetry (OSLD) or radiophotoluminescence dosimeters (RPLDs) and radiochromic gel dosimeters. These systems are discussed in brief in the next sections.

IV.1. RADIOCHROMIC FILMS

Radiochromic films can be used to perform measurements in HDR and LDR brachytherapy [271, 272], and in several cases where the use of the well-type chamber is not recommended for calibration. Radiochromic films have, for instance, been used for measurements of planar absorbed dose distributions in water and of the reference dose rates of beta sources [273, 274], for absorbed dose measurements in a water phantom of a miniature low energy X ray source [269], and for end-to-end dosimetric audits in solid water phantoms [275, 276].

IV.2. THERMOLUMINESCENT DOSIMETERS

Thermoluminescent dosimeters have long been used as dosimeters of choice in experimental brachytherapy dosimetry of low and high energy photon sources and have their role proved in the validation of the reference air kerma rate/air kerma strength, measurements of the source dose rate constant, the radial dose function and anisotropy function [79, 81, 82, 277–281].

The dose rate measurements were performed with TLDs in solid phantoms for intravascular brachytherapy beta sources [282] and ophthalmic (^{90}Sr, ^{106}Ru) applicators [283–285]. Other important applications included in vivo dosimetry, where TLDs were successfully used in gynaecological, prostate and skin brachytherapy treatments [286–288] and dosimetry audits in brachytherapy, comparing the measured and treatment planning system calculated absorbed dose to water [289, 290].

These dosimeters may have a relatively small active volume (e.g. $1 \times 1 \times 1$ mm^3 cube, $1 \times 1 \times 6$ mm^3 rod) to minimize the effect of high dose gradients across their volume while maintaining sufficient sensitivity for measurements. Accurate measurements with TLDs, typically performed in solid water phantoms with radiological properties comparable to water, require well-controlled irradiation conditions, careful handling and reading of the dosimeters. Their sensitivity highly depends on the applied heating and cooling cycles.

Due to the characteristics of the dose distribution around brachytherapy sources, especially in the immediate vicinity, the positional uncertainty can be considerable and very close tolerances in the measurements are required. Experimental and computational studies emphasized that the measured air kerma and TLD absorbed dose response of LiF:Mg,Ti (LiF:Mg,Cu,P) TLDs were not constant for a wide range of photon spectra [291, 292]. The absorbed dose sensitivity [82, 293–295] of the dosimeter can be divided into two main parts: the intrinsic energy dependence, which depends on the detector signal formation process and the linear energy transfer of the radiation (LET), and the energy dependence of the absorbed dose, which depends on the medium, the detector cross section, the self-attenuation and the volume averaging. To derive the absorbed dose energy dependence, either Monte Carlo simulations of the actual irradiation geometry or the use of cavity theory and knowledge of the absorption properties of water and the detector with additional perturbation correction factors are needed. If the measurement is performed in a medium other than the calibration medium, a detector response correction needs to also be applied to the phantom material. The correction is given by the ratio of dose to water at one point in water to dose to water at the same point in the phantom medium for the radiation quality used in the measurement [82].

Four methods for calibrating TLDs were investigated for use with an HDR ^{192}Ir brachytherapy source as the most suitable for audit purposes [296]. Three of the methods involved calibration with an HDR ^{192}Ir source, and for the fourth method a 6 MV photon beam was used. Calibration of TLDs in a phantom such as the one used for the auditing gave the most reliable results. The uncertainty of the method used to calibrate TLDs in 6 MV photon beams was the highest of all methods in the study, and the dose measurement results were consistently higher than those obtained using an HDR ^{192}Ir source.

IV.3. DIODES AND DIAMOND DETECTORS

Diodes and diamonds are solid state detectors, distinguished by the energy gap between the valence band and the conduction band. Diodes are based upon a small band gap. Diamonds have a large band gap separated by greater energy. This large band gap is the case for an insulator. The small size of both detectors is an advantage for brachytherapy applications, although they have not been extensively used [297–302].

Diamond is a carbon material and thus approximates tissue; however, diamonds are very expensive to purchase. Diamond detectors do not all perform similarly. The older style diamonds had great variability [303]. This variation was caused by the lack of purity or the amount of impurities in the crystal lattice. Impurities cause electron and hole traps and interfere with the radioconductivity signal. This is also true for diodes, although the solid state process is different. Quite promising results have been published about the newer micro-diamond detector [300–302, 304, 305].

Commercially available diodes are found to have up to 12% change in sensitivity with the angle of incidence of radiation [306]. It is important that any diode to be used be well characterized beforehand. There is also a significant energy dependence involved in the use of diodes; this becomes important when considering low energy brachytherapy sources. In addition, some diodes have a temperature effect; although the range of temperatures in normal use is not significant. A rule of thumb is about 0.3%/ °C variation in response [306]. The calibration of these diodes needs to be checked, especially if the accumulated dose reaches 100 Gy to 1 kGy.

IV.4. ALANINE

Measurements of the 3D dose distribution around brachytherapy sources is challenging, not only due to the steep dose gradients close to the source, but also the change of the energy spectra at different depths in the phantom medium. The requirements for a dosimetry system capable of addressing such challenges include the ability to measure a broad range of doses, a small detector size and, ideally, a detector response which is independent of radiation energy and dose rate. The dosimetry system also needs to make it possible to accurately position the detectors and the radiation source relative to one another in a repeatable way.

When the amino acid alanine is irradiated with ionizing radiation, stable free radicals are produced [307]. The number of free radicals is proportional to the absorbed radiation dose and alanine can therefore be used as a radiation dosimeter. Alanine is also near water equivalent which makes it appropriate for use in radiotherapy applications [308, 309]. Another advantage of alanine dosimeters is the non-destructive nature of their readout. The main issue with using alanine is the minimum measurable dose of around 5 Gy, which might require long exposure times.

Alanine powder and paraffin wax can be pressed into small cylindrical pellets with typical diameters and heights of a few millimetres. The solid alanine pellets can be placed in phantoms to measure absorbed dose at points close to brachytherapy sources.

The energy response of alanine in medium energy X rays with tube peak voltages ranging from 30 kV to 280 kV, which is relevant for LDR, PDR and HDR brachytherapy as well as eBT miniature X ray sources, have been investigated by different research teams [310–313]. Data on the response of alanine in ^{192}Ir photon radiation were published by Schaeken et al. [314]. Anton et al. [315] showed that the response of the alanine dosimeter to ^{192}Ir radiation relative to ^{60}Co radiation decreases from around 98% at 1 cm depth in water to 96% at 5 cm depth.

Alanine has been used for absolute dosimetry as part of a multicentre audit in the United Kingdom to evaluate HDR and PDR brachytherapy [162], for in vivo dosimetry to evaluate the urethra dose during HDR ^{192}Ir brachytherapy [316] and for measurements of absorbed dose around an HDR ^{192}Ir brachytherapy source which was placed inside an applicator [317].

Under the EMPIR project (18NRM02 PRISM-eBT, primary standards and traceable measurement methods for X-ray emitting eBT devices), various radiation detectors, including alanine, were characterized for dose distribution measurements close to some of the available low energy eBT miniature X-ray sources with or without applicators. The PRISM-eBT open access website can be found at http://www.ebt-empir.eu.

IV.5. OPTICALLY STIMULATED LUMINESCENCE DOSIMETERS (OSLDS)

The optically stimulated luminescence introduced in luminescence dating and promoted by retrospective and personal dosimetry has also been used in radiotherapy for clinical dosimetry measurements of high energy photon and electron beams using different types of readers and reading techniques. While several OSL materials such as BaO, KBr:Eu, Mg_2SiO_4:Tb or aluminium oxide doped with metal (Al_2O_3:Cr,Mg,Fe) have been investigated for dosimetry applications, Al_2O_3:C crystals are currently the only commercially available dosimeters.

The Al_2O_3:C crystals are converted into powder which is used to produce the plastic strip from which the small discs are extruded. The dosimeter nanoDotTM (Landauer Inc., Glenwood IL), for example, has a diameter of 4 mm and a thickness of only 0.2 mm. The other available form are strips that can be used for dose profile measurements in computed tomography. The advantages of these dosimeters for dose measurements around brachytherapy sources are their low thickness, which reduces the volume averaging effect in the steep dose gradients, high sensitivity, dose rate independence, reusability, easy readout and bleaching process, and stability against temperature and humidity variations. It is recommended to store the dosimeter in an adequate container so that room light will not affect its response.

In one study, an HDR ^{192}Ir brachytherapy source was used to test the dose linearity, dose rate dependence and angular response of OSLDs for potential in vivo HDR brachytherapy dosimetry [318]. The dosimetric performance of OSLDs evaluated in this study, together with the favourable practical properties of the dosimeters and the readout procedure, showed that OSLDs are feasible means for in vivo brachytherapy measurements.

The OSLDs characterized for use in the in vivo dosimetry of HDR ^{192}Ir brachytherapy [319] showed a dependence on the angle of incidence of a radiation field of a ^{192}Ir brachytherapy source on the detector surface. In addition, the authors proposed a calibration with a ^{192}Ir source to avoid the uncertainty in the different sensitivity of OSLDs to ^{192}Ir photons compared to 6 MV photon beams. A 10% increase in OSLD sensitivity, which depends on the changes in the photon spectrum, was found in measurements at 10 cm depth compared to measurements at 1 cm depth. The measurements performed were in good agreement with the TPS calculation when using an advanced model based dose calculation algorithm.

The feasibility of the OSLD-based ^{192}Ir HDR remote audit system, based on a mailable solid phantom preloaded with dosimeters and the developed methodology, showed in several experiments that the dose can be measured with uncertainties of ~2.5% ($k = 2$). The level of uncertainty was adequate to allow the audit acceptance limits of 5% to be established [320].

IV.6. RADIOPHOTOLUMINESCENCE DOSIMETERS

A silver activated phosphate glass is a basis for various types of radiophotoluminescence glass dosimeters due to its advantageous dosimetry properties and reproducible production. They have been thoroughly characterized and used in external beam radiotherapy large scale audits [321–325].

The commercially available RPLDs (e.g. GD-302M, Asahi Techno Glass Corporation, Japan) measure 1.5 mm in diameter and 12 mm in length and are encapsulated in a plastic capsule. Their effective atomic number is 12.04 and their density is 2.61 g/cm^3. The non-destructive readout process, in which only a small part of the signal is depleted, offers their reusability after annealing, in which luminescence centres are eliminated. The sensitive area of a dosimeter is 6 mm long. The automatic reader can read out up to 20 RPLDs in one short session.

Correction factors for absorbed dose energy dependence and intrinsic energy dependence were determined for absorbed dose to water measurements from a ^{192}Ir source using RPLDs calibrated in a 4 MV photon beam [326]. While the relative detector response for ^{192}Ir radiation and 4 MV photon beam did not vary significantly with distance from the source, the relative dose ratio of absorbed dose to water relative to average absorbed dose to RPLD for ^{192}Ir and 4 MV photons, corrected for the energy dependence of absorbed dose, decreased by about 20% for distances from 2 to 10 cm from the source.

The results of RPLD characterization measurements, including readout reproducibility, dose linearity and energy response, show that it has good radiation detection properties and is suitable for verification of the brachytherapy dose of HDR ^{192}Ir brachytherapy prostate treatments [327]. The results showed no significant energy dependence between ^{192}Ir and ^{60}Co sources for RPLDs after irradiation with both sources in the dose range of 100–700 cGy, and the calibration of dosimeters in the ^{60}Co beam quality was proposed. Radiophotoluminescence glass dosimetry and TLD measurements in the in-house made prostate phantom gave similar results. With multiple source positions the difference between the RPLD measurement results and the results calculated by TPS was within 5%.

Further applications of RPLDs include an in vivo dosimetry study for interstitial HDR ^{192}Ir head and neck brachytherapy [328]. The results of the study helped to determine the importance of dose prescription for achieving high reproducibility and avoidance of large hyperdose regions.

IV.7. GEL DOSIMETERS

Gel dosimeters have the attractive characteristic of being able to record the dose distribution in three dimensions. This characteristic is particularly significant in brachytherapy, where steep dose gradients around the source(s) are involved. Depending on their chemical formulations, gel dosimeters have in some cases shown to be close to water and tissue equivalent [329–331].

Gel dosimeters can, in general, be grouped in two main types, namely Fricke and polymer gels [332–335]. In Fricke gel dosimeters, radiation induces a change of ferrous (Fe^{2+}) into ferric (Fe^{3+}) ions whereas in polymer gel dosimeters, it induces a polymerization of the radiation sensitive chemical. In both cases, the changes induced by radiation are quantified and converted into an information of absorbed dose. Magnetic resonance imaging, optical or X-ray computed tomography and ultrasound have been investigated as possible methods to quantify these changes [332–334].

Even though they have already been studied for several decades, gel dosimeters have found a limited application in brachytherapy [336]. They have been used for three-dimensional dose determination around ^{192}Ir and ^{60}Co high dose rate sources. Their output was compared with the results of other dosimeters, Monte Carlo calculations and treatment planning systems based on AAPM TG-43 or model-based dose calculation algorithms [337–345]. Dose distributions close to LDR sources, such as ^{137}Cs, and beta emitters, such as ^{106}Ru and ^{90}Sr, were also quantified [333, 346–349].

IV.8. PLASTIC SCINTILLATORS

Having near tissue equivalence across a wide range of photon energies, sub-millimetre spatial resolution, and milli-second temporal capabilities such as being moved throughout a dose distribution or detecting a moving source, plastic scintillators have been used increasingly for brachytherapy dosimetry [184, 350]. Applications have been studied for measuring a single source within a phantom for acquiring reference dosimetry or for in vivo dosimetry for multi-dwell or multi-source brachytherapy implants [350–357].

Appendix V

THE AAPM TG-43 ALGORITHM FOR DOSE DISTRIBUTION CALCULATION IN BRACHYTHERAPY

The dose calculation formalism used for brachytherapy dose calculations is exhaustively provided in the AAPM TG-43 report and updates [79–83]. Since it is universally accepted as the standard, the algorithm for dose calculation is commonly cited in the literature as the 'TG-43 algorithm'. Model-based dose calculation algorithms were also recently proposed in brachytherapy; a description of these falls outside the scope of this code of practice and can be found elsewhere [358–361]. Only a brief description of the TG-43 algorithm is provided.

The TG-43 algorithm works in the approximation of a homogeneous infinite medium (i.e. full scatter conditions), with the delivered dose to a point of interest being the superposition of single source dose distributions to the same point, neglecting any inter-source or applicator attenuation effects [23]. These approximations are generally pertinent both for LDR and HDR clinical applications [358, 362]. The AKS is chosen in the TG-43 formalism to physically quantify the source strength. Information for the conversion from RAKR to AKS is provided in Section 3.1. The general TG-43 equations according to the 1D and 2D formalism are:

$$\dot{D}(r) = S_K \Lambda \frac{G_X(r,\theta_0)}{G_X(r_0,\theta_0)} g_X(r) \phi_{an}(r) \qquad \text{(1D formalism) (37)}$$

$$\dot{D}(r,\theta) = S_K \Lambda \frac{G_X(r,\theta)}{G_X(r_0,\theta_0)} g_X(r) F(r,\theta) \qquad \text{(2D formalism) (38)}$$

where S_K is the air kerma strength, Λ is the dose rate constant, G_X is the geometry function, g_X is the radial dose function, $\phi_{an}(r)$ and $F(r, \theta)$ are the 1D and 2D anisotropy functions, respectively, with the adopted polar coordinate system provided previously in Fig. 1. In the 1D formalism, a 1D isotropic point-source approximation is modelled and the dose depends only on the radial distance from the centre of the source. In the 2D formalism, a more complex 2D variation in dose distribution as a function also of polar angle relative to the source longitudinal axis is modelled. The latter formalism is more accurate, but it necessitates determination of the source transverse axis orientation from imaging studies. The variable X indicates whether the point-source (i.e. X = P) or line-source (i.e. X = L) model was selected. r_0 and θ_0 are the coordinates of a reference point P, with r_0 being 1 cm from the centre of the radioactive source and θ_0 specifying the transverse axis from the centre of the radioactive source (i.e. $\theta_0 = 90°$).

In general, it is important to note that both TG-43 algorithms are basically structured in two parts. The first part, which is common to both the 1D and 2D formalism, converts the AKS S_K to the dose rate in water at the reference point P:

$$\dot{D}(r_0,\theta_0) = S_K \Lambda \tag{39}$$

The second part allows one to calculate the dose rate in all the remaining points in water, starting from the dose rate at the reference point P:

$$\dot{D}(r) = \dot{D}(r_0,\theta_0)\frac{G_X(r,\theta_0)}{G_X(r_0,\theta_0)}g_X(r)\phi_{an}(r) \qquad \text{(1D formalism)} \tag{40}$$

$$\dot{D}(r) = \dot{D}(r_0,\theta_0)\frac{G_X(r,\theta_0)}{G_X(r_0,\theta_0)}g_X(r)F(r,\theta) \qquad \text{(2D formalism)} \tag{41}$$

V.1. DOSE RATE CONSTANT, Λ

The dose rate constant Λ is defined as:

$$\Lambda = \frac{\dot{D}(r_0,\theta_0)}{S_K} \tag{42}$$

and not only depends on the radionuclide, but also on the source model. Consensus data for clinical implementation of Λ for the different source models can be found elsewhere [23, 81], and are obtained and validated combining the information provided by specific Monte Carlo calculations and experimental measurements published in peer-reviewed scientific journals.

V.2. GEOMETRY FUNCTION G_X

The geometry function $G_X(r, \theta)$ accounts for the inverse square law and does not consider radiation absorption and scattering by the traversed means. In the line-source approximation model, G takes also into account the approximate spatial distribution of radioactivity within the active core of the source. Geometry functions according to the point-source and line-source approximations are:

$$G_P(r,\theta) = r^{-2} \quad \text{point-source approximation} \tag{43}$$

$$G_L(r,\theta) = \begin{cases} \dfrac{\beta}{Lr\sin\theta} & \text{if } \theta \neq 0^\circ \\[2ex] \left(r^2 - \dfrac{L^2}{4}\right)^{-1} & \text{if } \theta = 0^\circ \end{cases} \quad \text{line-source approximation} \tag{44}$$

To obtain unity in P, in the TG-43 algorithm $G_X(r, \theta)$ is normalized by $G_X(r_0, \theta_0)$.

V.3. RADIAL DOSE FUNCTION G_X

The radial dose function $g_X(r)$ accounts for the dose rate change due to photon scattering and absorption along the transversal axis through the centre of the source, excluding the dose geometrical fall-off effects modelled with $G_X(r, \theta)$. Overall, $g_X(r)$ is defined as:

$$g_X(r) = \frac{\dot{D}(r,\theta_0)}{\dot{D}(r_0,\theta_0)} \frac{G_X(r_0,\theta_0)}{G_X(r,\theta_0)} \tag{45}$$

The value of g_X is unity in r_0. In general, to define consensus data, Monte Carlo and experimental results published in peer-reviewed scientific journals were compared.

V.4. ANISOTROPY FUNCTIONS Φ_{an} AND F

The 1D anisotropy function $\Phi_{an}(r)$ is defined at a given distance r as "the ratio of the solid angle-weighted dose rate, averaged over the entire 4π steradian space, to the dose rate at the same distance r on the transverse plane" [81]:

$$\phi_{an}(r) = \frac{\int_0^\pi \dot{D}(r,\theta)\sin(\theta)d\theta}{2\dot{D}(r,\theta_0)} \tag{46}$$

Since the dose might in fact be different over the entire steradian space at a certain distance r, $\Phi_{an}(r)$ approximates this difference by calculating the average value.

On the contrary, "the 2D anisotropy function describes the variation in dose as a function of polar angle θ relative to the transverse plane" [81]. For sources different than point-sources, this angular variation is mainly due to self-filtration, scattering of photons within the source and oblique filtration of primary photons through the source encapsulation. $F(r, \theta)$ is defined as:

$$F(r,\theta) = \frac{\dot{D}(r,\theta)}{\dot{D}(r,\theta_0)} \frac{G_L(r,\theta_0)}{G_L(r,\theta)} \tag{47}$$

and equals unity along the transverse plane (i.e. $F(r, \theta_0) = 1$).

Appendix VI

EXPRESSION OF UNCERTAINTIES[6]

VI.1. INTRODUCTION

The aim of any measurement is to obtain the value of a parameter or quantity, generally termed measurand. The uncertainty associated with a measurement is a parameter that characterizes the dispersion of the values "that could reasonably be attributed to the measurand". This parameter is normally an estimated standard deviation. An uncertainty, therefore, has no known sign and is usually assumed to be symmetrical. It is a measure of our lack of exact knowledge, after all recognized 'systematic' effects have been eliminated by applying appropriate corrections.

The ISO Guide on the Expression of Uncertainty in Measurement [167] gives definitions and describes methods of evaluating and reporting uncertainties. It presents a consensus on how the uncertainty in measurement is generally treated. The guide suggests using Type A and Type B uncertainties based on the method used to evaluate the uncertainty. Statistical methods are used to evaluate Type A uncertainties as opposed to Type B uncertainties which are determined by other means. In this code of practice, the ISO guide is followed, and it is advised to consult it for details when needed.

VI.2. MEAN VALUE OF MEASUREMENT

In a series of n measurements, with observed values x_i, the best estimate of the quantity x is usually given by the *arithmetic mean* value:

$$\bar{x} = \frac{1}{n} \sum_{i=1}^{n} x_i \tag{48}$$

The scatter of the n measured values x_i, around their mean $\bar{x}$ can be characterized by the *standard deviation*:

$$s(x_i) = \sqrt{\frac{1}{n-1} \sum_{i=1}^{n} (x_i - \bar{x})^2} \tag{49}$$

---

[6] This appendix is mainly taken from appendix I of the TRS-457 [41].

and the quantity $s^2(x_i)$ is called the empirical variance of a single measurement, based on a sample of size n.

The standard deviation of the mean value, written as $s(\bar{x})$, can be calculated according to:

$$s(\bar{x}) = \frac{1}{\sqrt{n}} s(x_i) \tag{50}$$

An alternative way to estimate $s(\bar{x})$ would be based on the outcome of several groups of measurements. If they are all of the same size, the formulas given above can still be used, provided that x_i is now taken as the mean of group i and $\bar{x}$ is the overall mean (or mean of the means) of the n groups. For groups of different size, *statistical weights* would have to be used. This second approach may often be preferable, but it usually requires a larger number of measurements. A discussion of how much the two results of $s(\bar{x})$ may differ from each other is beyond this elementary presentation.

VI.3. TYPE A STANDARD UNCERTAINTY

The *standard uncertainty of Type A*, denoted by u_A, is described by the standard deviation of the mean value of statistically independent observations, or

$$u_A = s(\bar{x}) \tag{51}$$

This equation shows that a Type A uncertainty of the measurement of a quantity can, in principle, always be reduced by increasing the number n of individual readings. It must be noted that the reliability of a Type A uncertainty estimation according to Eq. (51) has to be considered for the low number of measurements ($n < 10$). Other means of estimations, such as the t-distribution may be considered. If several measurement techniques are available, preference will be given to the one which produces the least scatter of the results (i.e. which has the smallest standard deviation $s(x_i)$, but in practice the possibilities for reduction are often limited). One example is the measurement of ionization currents that are of the same order as the leakage currents, which may also be variable. In order to arrive at an acceptable uncertainty of the result, it is then necessary to take many more readings than would normally be needed if the ionization currents were much higher than the leakage currents.

The Type A standard uncertainty is obtained by the usual statistical analysis of repeated measurements. It is normally found that the reproducibility of each model of dosimeter is essentially the same from one instrument to the next.

Thus, if the Type A standard uncertainty of an air-kerma-rate measurement is determined for one kind of dosimeter, the same value can generally be used for other instruments of that same model, used under the same conditions.

VI.4. TYPE B STANDARD UNCERTAINTY

There are many sources of measurement uncertainty that cannot be estimated by repeated measurements. These are called Type B uncertainties. These include not only unknown, although suspected, influences on the measurement process, but also little known effects of influence quantities (mechanical deform of an ionization chamber due to temperature and humidity), application of correction factors or physical data taken from the literature, experience from previous measurements, manufacturer's specifications, etc. A calibration uncertainty, even if derived from Type A components, becomes a Type B uncertainty when using the calibrated instrument.

In the CIPM method of characterizing uncertainties, the Type B uncertainties have to be estimated so that they correspond to standard deviations; they are called Type B standard uncertainties. Some experimenters claim that they can directly estimate this type of uncertainty, while others prefer to use, as an intermediate step, some type of limit. It is often helpful to assume that these uncertainties have a probability distribution which corresponds to some easily recognizable shape.

If, for example, one is 'fairly sure' of that limit, L, it can be considered to correspond approximately to a 95% confidence limit, whereas, if one is 'almost certain', it may be taken to correspond approximately to a 99% confidence limit. Thus, the Type B standard uncertainty, u_B, can be obtained from the equation:

$$u_B = \frac{L}{k} \tag{52}$$

where $k = 2$ if one is fairly certain, and $k = 3$ if one is quite certain of the estimated limits $\pm L$. These relations correspond to the properties of a Gaussian distribution and it is usually not worthwhile to apply divisors other than 2 or 3 because of the approximate nature of the estimation [363, 364].

It is sometimes assumed that Type B uncertainties can be described by a rectangular probability density function; in other words, that they have equal probability anywhere within the given maximum limits −M and +M and their

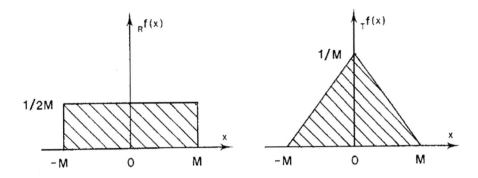

FIG. 7. Rectangular $_Rf(x)$ and triangular $_Tf(x)$ probability density functions used in some cases to model unknown distributions.

probability is zero outside of these limits (see Fig. 7). It can be shown that with this assumption the Type B standard uncertainty u_B is given by:

$$u_B = \frac{M}{\sqrt{3}}$$

(53)

Alternatively, if the assumed distribution is triangular and with the same limits (see Fig. 7), the standard uncertainty can be expressed as:

$$u_B = \frac{M}{\sqrt{6}}$$

(54)

There are thus no rigid rules for estimating Type B standard uncertainties. The best knowledge and experience to estimate them is needed. In practice, usually very little is known about the uncertainty distribution and its choice is somewhat arbitrary. As most of uncertainty sources have Gaussian distribution, it is preferable to use this model when the exact shape of the distribution is unknown. But this applies only to situations when the uncertainty is not a dominant part of the overall uncertainty. In such a case, the uncertainty distribution can be reliably estimated [363]. The proper use of available information for evaluation of a Type B uncertainty requires a good general knowledge and experience.

VI.5. COMBINED UNCERTAINTIES AND EXPANDED UNCERTAINTIES

Type A and Type B uncertainties are both estimated standard deviations, so they are combined using the statistical rules for combining variances (which are squares of standard deviations). If u_A and u_B are the Type A and the Type B standard uncertainties of a quantity, respectively, the combined standard uncertainty u_C of that quantity is:

$$u_C = \sqrt{u_A^2 + u_B^2} \qquad (55)$$

This equation is strictly valid only provided that the uncertainty sources are not correlated. The correlation terms have to be considered in the expression for u_C if some of the uncertainties are not completely independent. An example is a difference or ratio of two measurements made by the same instrument. Details of a correlation treatment can be found elsewhere [167].

The combined standard uncertainty still has the character of a standard deviation. If, in addition, it is believed to have a Gaussian probability density, then the standard deviation corresponds to a confidence limit of about 68%. Therefore, it is often felt desirable to multiply the combined standard uncertainty by a suitable factor, called the coverage factor, k, to yield an expanded uncertainty:

$$U = k u_C \qquad (56)$$

Suitable values of the coverage factor are $k = 2$ or 3, corresponding to confidence limits of about 95% or 99%. The approximate nature of uncertainty estimates, in particular for Type B, makes it doubtful that more than one significant figure is ever justified in choosing the coverage factor. In any case, it is important to clearly indicate the numerical value taken for the coverage factor. The expanded uncertainty is also known under the name 'overall uncertainty'.

VI.6. PROPAGATION OF UNCERTAINTIES

The expression 'propagation of errors' was part of the statistical terminology before it became customary to distinguish between errors and uncertainties, and it is still occasionally used. In order to be consistent with the present terminology, it is preferable to talk about the propagation of uncertainties.

Let us first consider a practical example. The calibration coefficient determined by a given calibration laboratory is not only based on various measurements performed at the laboratory, but also on correction factors and physical constants, as well as on a beam calibration traceable to a secondary laboratory, and ultimately, to a primary laboratory. All these numerical values contain uncertainties, and they combine to a given final uncertainty in the calibration coefficient. This situation can be represented in more general terms by considering a variable y which is a function of a number of variables $x_1, x_2, x_3, \dots$ This can be written in the form:

$$y = f(x_1, x_2, x_3, \dots) \qquad (57)$$

In many practical cases, the influence quantities $x_1, x_2, x_3, \dots$ are independent of each other. Then $u(y)$ can be calculated by the simple formula:

$$u(y) \cong \sqrt{\left(\frac{\partial f}{\partial x_1}\right)^2 u^2(x_1) + \left(\frac{\partial f}{\partial x_2}\right)^2 u^2(x_2) + \left(\frac{\partial f}{\partial x_3}\right)^2 u^2(x_3) + \cdots} \qquad (58)$$

Two special cases should be mentioned, in particular since they are of great practical importance and cover most of the usual situations.

If the functional dependence is linear (i.e. for sums or differences) the following equation applies:

$$y = c_1 x_1 + c_2 x_2 + c_3 x_3 + \cdots \qquad (59)$$

where

$$c_i = \frac{\partial y}{\partial x_1} \qquad (60)$$

and c_i is the sensitivity coefficient for the input quantity x_i. Then, the uncertainty on y is:

$$u(y) = \sqrt{c_1^2 u^2(x_1) + c_2^2 u^2(x_2) + c_3^2 u^2(x_3) + \cdots} \qquad (61)$$

Thus, if independent variables are added (or subtracted), the variances also add. In other words, the uncertainty of the sum is obtained by adding in quadrature the 'weighted' uncertainties of the independent variables, where the 'weights' are the squares of the coefficients $c_1, c_2, c_3, \dots$ ('adding in quadrature' means taking the square root of the sum of the squares).

The other special case concerns a product (or ratio) of independent variables. The functional dependence then is:

$$y = x_1^{\alpha} x_2^{\beta} x_3^{\gamma} \ldots \tag{62}$$

where the exponents α, β, γ, ... are constants. In this case, the following expression for the relative uncertainty on y is obtained from Eq. (62):

$$r(y) = \sqrt{\alpha^2 r^2(x_1) + \beta^2 r^2(x_2) + \gamma^2 r^2(x_3) + \cdots} \tag{63}$$

where

$$r(x_i) = \frac{u(x_i)}{|x_i|} \tag{64}$$

is the relative uncertainty of x_i.

Thus, for a product (or ratio) of independent variables, the relative weighted variances add, where the weights are the squares of the exponents α, β, γ, ... A very common case is that of a ratio, $y = x_1 / x_2$, where the quantities x_1 and x_2 contain measurements and correction factors. From Eq. (63) the relative variance on y is equal to the quadratic sum of the relative uncertainties on x_1 and x_2.

The foregoing discussion applies to Type A, Type B, and combined standard uncertainties, all of which are estimated as to correspond to standard deviations. The rules for propagation of uncertainty also apply to expanded uncertainties, provided that everywhere the same coverage factor k has been used. The uncertainty on published data is generally in terms of an expanded uncertainty, or some equivalent terminology. This has then to be converted into a standard deviation, before using it to calculate an uncertainty. If no coverage factor is stated, it may be assumed to have the value $k = 2$.

It is preferable to tabulate both Type A and Type B standard uncertainties separately. This makes possible later changes easier to perform.

REFERENCES

[1]   INTERNATIONAL ATOMIC ENERGY AGENCY, Radiation Oncology Physics: A Handbook for Teachers and Students, STI/PUB/1196, IAEA, Vienna (2005).

[2]   PARK, C.C., et al., American Society for Therapeutic Radiology and Oncology (ASTRO) Emerging Technology Committee report on electronic brachytherapy, Int. J. Radiat. Oncol. Biol. Phys. **76** (2010) 963–972.

[3]   THOMADSEN, B.R., et al., Electronic intracavitary brachytherapy quality management based on risk analysis: The report of AAPM TG 182, Med. Phys. (2019) e65–e91.

[4]   INTERNATIONAL AGENCY FOR RESEARCH ON CANCER – WORLD HEALTH ORGANIZATION (2018),
        https://gco.iarc.fr/today/home

[5]   HAN, K., MILOSEVIC, M., FYLES, A., PINTILIE, M., VISWANATHAN, A.N., Trends in the utilization of brachytherapy in cervical cancer in the United States, Int. J. Radiat. Oncol. Biol. Phys. **87** (2013) 111–119.

[6]   TANDERUP, K., EIFEL, P.J., YASHAR, C.M., POTTER, R., GRIGSBY, P.W., Curative radiation therapy for locally advanced cervical cancer: brachytherapy is NOT optional, Int. J. Radiat. Oncol. Biol. Phys. **88** (2014) 537–539.

[7]   CHARGARI, C., et al., Brachytherapy: An overview for clinicians, CA-Cancer J. Clin. **69** (2019) 386–401.

[8]   KISHAN, A.U., et al., Radical Prostatectomy, External Beam Radiotherapy, or External Beam Radiotherapy With Brachytherapy Boost and Disease Progression and Mortality in Patients With Gleason Score 9-10 Prostate Cancer, JAMA **319** (2018) 896–905.

[9]   INTERNATIONAL ATOMIC ENERGY AGENCY, Directory of Radiotherapy Centres (DIRAC) (2020),
        https://dirac.iaea.org/

[10]   INTERNATIONAL ATOMIC ENERGY AGENCY, The Transition from 2-D Brachytherapy to 3-D High Dose Rate Brachytherapy, Human Health Reports No. 12, IAEA, Vienna (2015).

[11]   INTERNATIONAL ATOMIC ENERGY AGENCY, Lessons Learned from Accidental Exposures in Radiotherapy, Safety Reports Series No. 17, IAEA, Vienna (2000).

[12]   INTERNATIONAL COMMISSION ON RADIOLOGICAL PROTECTION, Prevention of High-Dose-Rate Brachytherapy Accidents, ICRP Publication No. 97, ICRP Publication No. 97 — Annals of the ICRP 35 (2), 2005, ICRP, Oxford (2005).

[13]   DEUFEL, C.L., et al., Patient safety is improved with an incident learning system— Clinical evidence in brachytherapy, Radiother. Oncol. **125** (2017) 94–100.

[14]   INTERNATIONAL ATOMIC ENERGY AGENCY, Applying radiation safety standards in radiotherapy, Safety Report Series No. 38, IAEA, Vienna (2006).

[15]   UNITED STATES NUCLEAR REGULATORY COMMISSION, Reportable Medical Events Involving Treatment Delivery Errors Caused by Confusion of Units for the Specification of Brachytherapy Sources, U.S. Nuclear Regulatory Commission, Office of Federal and State Materials and Environmental Management Programs, Washington, DC (2009).

[16] INTERNATIONAL ATOMIC ENERGY AGENCY, Calibration of Brachytherapy Sources: Guidelines on Standardized Procedures for the Calibration of Brachytherapy Sources at Secondary Standard Dosimetry Laboratories (SSDLs) and Hospitals, IAEA-TECDOC-1079, IAEA, Vienna (1999).

[17] VAN DER MERWE, D., et al., Accuracy requirements and uncertainties in radiotherapy: a report of the International Atomic Energy Agency, Acta Oncol. **56** (2017) 1–6.

[18] INTERNATIONAL ATOMIC ENERGY AGENCY, Accuracy Requirements and Uncertainties in Radiotherapy, Human Health Series No. 31, IAEA, Vienna (2016).

[19] KIRISITS, C., et al., Review of clinical brachytherapy uncertainties: Analysis guidelines of GEC-ESTRO and the AAPM, Radiother. Oncol. **110** (2014) 199–212.

[20] BALGOBIND, B.V., et al., A review of the clinical experience in pulsed dose rate brachytherapy, Br. J. Radiol. **88** (2015) 20150310.

[21] PAPAGIANNIS, P., et al., Monte Carlo dosimetry of ^{60}Co HDR brachytherapy sources, Med. Phys. **30** (2003) 712–721.

[22] GRANERO, D., PEREZ-CALATAYUD, J., BALLESTER, F., Technical note: Dosimetric study of a new Co-60 source used in brachytherapy, Med. Phys. **34** (2007) 3485–3488.

[23] PEREZ-CALATAYUD, J., et al., Dose calculation for photon-emitting brachytherapy sources with average energy higher than 50 keV: report of the AAPM and ESTRO, Med. Phys. **39** (2012) 2904–2929.

[24] EATON, D.J., Electronic brachytherapy—current status and future directions, Br. J. Radiol. **88** (2015) 20150002.

[25] BUTLER, W.M., et al., Third-party brachytherapy source calibrations and physicist responsibilities: report of the AAPM Low Energy Brachytherapy Source Calibration Working Group, Med. Phys. **35** (2008) 3860–3865.

[26] CHIU-TSAO, S.T., SCHAART, D.R., SOARES, C.G., NATH, R., Dose calculation formalisms and consensus dosimetry parameters for intravascular brachytherapy dosimetry: Recommendations of the AAPM Therapy Physics Committee Task Group No. 149, Med. Phys. **34** (2007) 4126–4157.

[27] AMERICAN BRACHYTHERAPY SOCIETY - OPHTHALMIC ONCOLOGY TASK FORCE, ABS — OOTF COMMITTEE, The American Brachytherapy Society consensus guidelines for plaque brachytherapy of uveal melanoma and retinoblastoma, Brachytherapy **13** (2014) 1–14.

[28] THOMSON, R.M., et al., AAPM recommendations on medical physics practices for ocular plaque brachytherapy: Report of task group 221, Med. Phys. **47** (2020) e92–e124.

[29] GUINOT, J.L., et al., GEC-ESTRO ACROP recommendations in skin brachytherapy, Radiother. Oncol. **126** (2018) 377–385.

[30] AKULINICHEV, S., DERZHIEV, V., KRAVCHUK, L., Ytterbium Sources for brachytherapy, Radiother. Oncol. **102** (2012) S73–S74.

[31] KRISHNAMURTHY, D., CUNHA, J., HSU, I., WEINBERG, V., POULIOT, J., Dosimetric Comparison of iridium-192, ytterbium-169, and thulium-170 sources for HDR prostate brachytherapy, Med. Phys. **37** (2010) 3196.

[32] LEONARD, K.L., DIPETRILLO, T.A., MUNRO, J.J., WAZER, D.E., A novel ytterbium-169 brachytherapy source and delivery system for use in conjunction with minimally invasive wedge resection of early-stage lung cancer, Brachytherapy **10** (2011) 163–169.

[33] ENGER, S.A., LUNDQVIST, H., D'AMOURS, M., BEAULIEU, L., Exploring Co-57 as a new isotope for brachytherapy applications, Med. Phys. **39** (2012) 2342–2345.

[34] KRISHNAMURTHY, D., WEINBERG, V., CUNHA, J.A.M., HSU, I.C., POULIOT, J., Comparison of high-dose rate prostate brachytherapy dose distributions with iridium-192, ytterbium-169, and thulium-170 sources, Brachytherapy **10** (2011) 461–465.

[35] ENGER, S.A., FISHER, D.R., FLYNN, R.T., Gadolinium-153 as a brachytherapy isotope, Phys. Med. Biol. **58** (2013) 957–964.

[36] INTERNATIONAL ATOMIC ENERGY AGENCY, Calibration of Photon and Beta Ray Sources Used in Brachytherapy: Guidelines on Standardized Procedures at Secondary Standards Doşimetry Laboratories (SSDLs) and Hospitals, IAEA-TECDOC-1274, IAEA, Vienna (2002).

[37] NATH, R., et al., Code of practice for brachytherapy physics: report of the AAPM Radiation Therapy Committee Task Group No. 56. American Association of Physicists in Medicine, Med. Phys. **24** (1997) 1557–1598.

[38] GOETSCH, S.J., ATTIX, F.H., DeWERD, L.A., THOMADSEN, B.R., A new re-entrant ionization chamber for the calibration of iridium-192 high dose rate sources, Int. J. Radiat. Oncol. Biol. Phys. **24** (1992) 167–170.

[39] MITCH, M.G., ZIMMERMAN, B.E., LAMPERTI, P.J., SELTZER, S.M., COURSEY, B.M., Well-ionization chamber response relative to NIST air-kerma strength standard for prostate brachytherapy seeds, Med. Phys. **27** (2000) 2293–2296.

[40] INTERNATIONAL ATOMIC ENERGY AGENCY, Absorbed Dose Determination in External Beam Radiotherapy: An International Code of Practice for Dosimetry Based on Standards of Absorbed Dose to Water, Technical Reports Series No. 398, IAEA, Vienna (2000).

[41] INTERNATIONAL ATOMIC ENERGY AGENCY, Dosimetry in Diagnostic Radiology: An International Code of Practice, Technical Reports Series No. 457, IAEA, Vienna (2007).

[42] EUROPEAN COMMISSION, et al., Radiation protection and safety of radiation sources: International basic safety standards: General Safety Requirements Part 3, No. GSR Part 3, IAEA, Vienna (2014).

[43] INTERNATIONAL COMMISSION ON RADIATION UNITS AND MEASUREMENTS, Dose and Volume Specification for Reporting Intracavitary Therapy in Gynecology, ICRU Report 38, Bethesda, MD (1985).

[44] KIRISITS, C., et al., Basic treatment planning parameters for a ^{90}Sr / ^{90}Y source train used in endovascular brachytherapy, Z. Med. Phys. **14** (2004) 159–167.

[45] SIDAWY, A.N., WEISWASSER, J.M., WAKSMAN, R., Peripheral vascular brachytherapy, J. Vasc. Surg. **35** (2002) 1041–1047.

[46] TRIPURANENI, P., GIAP, H., JANI, S., Endovascular brachytherapy for peripheral vascular disease, Semin. Radiat. Oncol. **9** (1999) 190–202.

[47]   KIRWAN, J.F., CONSTABLE, P.H., MURDOCH, I.E., KHAW, P.T., Beta irradiation: new uses for an old treatment: a review, Eye (Lond) **17** (2003) 207–215.

[48]   PE'ER, J., Ruthenium-106 brachytherapy, Dev. Ophthalmol. **49** (2012) 27–40.

[49]   NEAL, A.J., IRWIN, C., HOPE-STONE, H.F., The role of strontium-90 beta irradiation in the management of pterygium, Clin. Oncol. (R. Coll. Radio.l). **3** (1991) 105–109.

[50]   NISHIMURA, Y., et al., Long-term results of fractionated strontium-90 radiation therapy for pterygia, Int. J. Radiat. Oncol. Biol. Phys. **46** (2000) 137–141.

[51]   VIANI, G.A., STEFANO, E.J., DE FENDI, L.I., FONSECA, E.C., Long-term results and prognostic factors of fractionated strontium-90 eye applicator for pterygium, Int. J. Radiat. Oncol. Biol. Phys. **72** (2008) 1174–1179.

[52]   HOLMES, S.M., MICKA, J.A., DeWERD, L.A., Ophthalmic applicators: An overview of calibrations following the change to SI units, Med. Phys. **36** (2009) 1473–1477.

[53]   HANSEN, J.B., CULBERSON, W.S., DeWERD, L.A., A convex windowless extrapolation chamber to measure surface dose rate from ^{106}Ru/^{106}Rh episcleral plaques, Med. Phys. **46** (2019) 2430–2443.

[54]   KOLLAARD, R.P., et al., Recommendations on detectors and quality control procedures for brachytherapy beta sources, Radiother. Oncol. **78** (2006) 223–229.

[55]   KRAUSE, F., MOLLER, M., RISSKE, F., SIEBERT, F.A., Dosimetry of ruthenium-106 ophthalmic applicators with thin layer thermoluminescence dosimeters - Clinical quality control, Z. Med. Phys. **30** (2020) 142–147.

[56]   SOARES, C.G., et al., Dosimetry of beta-ray ophthalmic applicators: comparison of different measurement methods, Med. Phys. **28** (2001) 1373–1384.

[57]   NATH, R., et al., Guidelines by the AAPM and GEC-ESTRO on the use of innovative brachytherapy devices and applications: Report of Task Group 167, Med. Phys. **43** (2016) 3178–3205.

[58]   FLYNN, R.T., et al., Efficient Yb-169 high-dose-rate brachytherapy source production using reactivation, Med. Phys. **46** (2019) 2935–2943.

[59]   VANDAMME, J.J., CULBERSON, W.S., DeWERD, L.A., MICKA, J.A., Air-kerma strength determination of a ^{169}Yb high dose rate brachytherapy source, Med. Phys. **35** (2008) 3935–3942.

[60]   LIN, L., PATEL, R.R., THOMADSEN, B.R., HENDERSON, D.L., The use of directional interstitial sources to improve dosimetry in breast brachytherapy, Med. Phys. **35** (2008) 240–247.

[61]   CHASWAL, V., THOMADSEN, B.R., HENDERSON, D.L., Development of an adjoint sensitivity field-based treatment-planning technique for the use of newly designed directional LDR sources in brachytherapy, Phys. Med. Biol. **57** (2012) 963–982.

[62]   AIMA, M., REED, J.L., DeWERD, L.A., CULBERSON, W.S., Air-kerma strength determination of a new directional ^{103}Pd source, Med. Phys. **42** (2015) 7144–7152.

[63]   RIVARD, M.J., A directional ^{103}Pd brachytherapy device: Dosimetric characterization and practical aspects for clinical use, Brachytherapy 16 (2017) 421–432.

[64]   AIMA, M., DeWERD, L.A., MITCH, M.G., HAMMER, C.G., CULBERSON, W.S., Dosimetric characterization of a new directional low-dose rate brachytherapy source, Med. Phys. **45** (2018) 3848–3860.

[65] COHEN, G.N., et al., Intraoperative implantation of a mesh of directional palladium sources (CivaSheet): Dosimetry verification, clinical commissioning, dose specification, and preliminary experience, Brachytherapy **16** (2017) 1257–1264.

[66] SIOSHANSI, S., et al., Dose modeling of noninvasive image-guided breast brachytherapy in comparison to electron beam boost and three-dimensional conformal accelerated partial breast irradiation, Int. J. Radiat. Oncol. Biol. Phys. **80** (2011) 410–416.

[67] LEONARD, K.L., et al., Prescription dose evaluation for APBI with noninvasive image-guided breast brachytherapy using equivalent uniform dose, Brachytherapy **14** (2015) 496–501.

[68] RIVARD, M.J., MELHUS, C.S., WAZER, D.E., BRICAULT, R.J., Jr., Dosimetric characterization of round HDR 192Ir accuboost applicators for breast brachytherapy, Med. Phys. **36** (2009) 5027–5032.

[69] YANG, W., et al., Rotating-shield brachytherapy for cervical cancer, Phys. Med. Biol. **58** (2013) 3931–3941.

[70] EBERT, M.A., Possibilities for intensity-modulated brachytherapy: technical limitations on the use of non-isotropic sources, Phys. Med. Biol. **47** (2002) 2495–2509.

[71] EBERT, M.A., Potential dose-conformity advantages with multi-source intensity-modulated brachytherapy (IMBT), Australas. Phys. Eng. Sci. Med. **29** (2006) 165–171.

[72] WEBSTER, M.J., et al., HDR brachytherapy of rectal cancer using a novel grooved-shielding applicator design, Med. Phys. **40** (2013) 091704.

[73] WEBSTER, M.J., et al., Dynamic modulated brachytherapy (DMBT) for rectal cancer, Med. Phys. **40** (2013) 011718.

[74] ADAMS, Q.E., et al., Interstitial rotating shield brachytherapy for prostate cancer, Med. Phys. **41** (2014) 051703.

[75] FAMULARI, G., URLICH, T., ARMSTRONG, A., ENGER, S.A., Practical aspects of 153Gd as a radioactive source for use in brachytherapy, Appl. Radiat. Isot. **130** (2017) 131–139.

[76] SARFEHNIA, A., STEWART, K., SEUNTJENS, J., An absorbed dose to water standard for HDR 192Ir brachytherapy sources based on water calorimetry: numerical and experimental proof-of-principle, Med. Phys. **34** (2007) 4957–4961.

[77] SANDER, T., Air kerma and absorbed dose standards for reference dosimetry in brachytherapy, Br. J. Radiol. **87** (2014) 20140176.

[78] SARFEHNIA, A., SEUNTJENS, J., Development of a water calorimetry-based standard for absorbed dose to water in HDR 192Ir brachytherapy, Med. Phys. **37** (2010) 1914–1923.

[79] NATH, R., et al., Dosimetry of interstitial brachytherapy sources: Recommendations of the AAPM Radiation Therapy Committee Task Group No. 43. American Association of Physicists in Medicine, Med. Phys. **22** (1995) 209–234.

[80] RIVARD, M.J., et al., Supplement to the 2004 update of the AAPM Task Group No. 43 Report, Med. Phys. **34** (2007) 2187–2205.

[81] RIVARD, M.J., et al., Update of AAPM Task Group No. 43 Report: A revised AAPM protocol for brachytherapy dose calculations, Med. Phys. **31** (2004) 633–674.

[82] RIVARD, M.J., et al., Supplement 2 for the 2004 update of the AAPM Task Group No. 43 Report: Joint recommendations by the AAPM and GEC-ESTRO, Med. Phys. **44** (2017) e297–e338.

[83] WILLIAMSON, J.F., et al., Recommendations of the American Association of Physicists in Medicine regarding the impact of implementing the 2004 task group 43 report on dose specification for ^{103}Pd and ^{125}I interstitial brachytherapy, Med. Phys. **32** (2005) 1424–1439.

[84] THOMADSEN, B., et al., Analysis of treatment delivery errors in brachytherapy using formal risk analysis techniques, Int. J. Radiat. Oncol. Biol. Phys. **57** (2003) 1492–1508.

[85] INTERNATIONAL COMMISSION ON RADIATION UNITS AND MEASUREMENTS, Dose and Volume Specification for Reporting Interstitial Therapy, ICRU Report 58, Bethesda, MD (1997).

[86] INTERNATIONAL COMMISSION ON RADIATION UNITS AND MEASUREMENTS, Prescribing, Recording, and Reporting Brachytherapy for Cancer of the Cervix, ICRU Report 89, (2013).

[87] INTERNATIONAL COMMISSION ON RADIATION UNITS AND MEASUREMENTS, Dosimetry of Beta Rays and Low-energy Photons for Brachytherapy with Sealed Sources, ICRU Report 72, (2004).

[88] PEREZ-CALATAYUD, J., et al., GEC-ESTRO ACROP recommendations on calibration and traceability of LE-LDR photon-emitting brachytherapy sources at the hospital level, Radiother. Oncol. **135** (2019) 120–129.

[89] JOINT AAPM/IROC HOUSTON REGISTRY OF BRACHYTHERAPY SOURCES MEETING THE AAPM DOSIMETRIC PREREQUISITES (2022), http://rpc.mdanderson.org/RPC/BrachySeeds/Source_Registry.htm

[90] TAYLOR, R.E.P., ROGERS, D.W.O., The CLRP TG-43 Parameter Database for Brachytherapy (2022), https://physics.carleton.ca/clrp/seed_database

[91] UNIVERSITY OF VALENCIA (UVEG) VALENCIA SPAIN, Dosimetry Parameters for Source Models used in Brachytherapy (2022), https://www.uv.es/braphyqs/

[92] HOLMBERG, O., Accident prevention in radiotherapy, Biomed. Imaging Interv. J. **3** (2007) e27.

[93] VENSELAAR, J., BALTAS, D., "Brachytherapy Physics: Sources and Dosimetry", The GEC ESTRO Handbook of Brachytherapy, ESTRO, Brussels (2014) 3–23.

[94] BE, M.M., et al., Table of Radionuclides Commissariat a l'Energie Atomique, Paris (1999).

[95] BE, M.M., et al., Monographie BIPM-5, Table of Radinuclides (Vol. 3 – A = 3 to 244), Bureau International des Poids et Mesures, (2006).

[96] BE, M.M., et al., Monographie BIPM-5, Table of Radionuclides (Vol. 6 – A = 22 to 242), Bureau International des Poids et Mesures, (2011).

[97] BE, M.M., et al., Monographie BIPM-5, Table of Radionuclides (Vol. 8 – A = 41 to 198), Bureau International des Poids et Mesures, (2016).

[98]   LABORATOIRE NATIONAL HENRI BECQUEREL FRANCE, Decay Data Evaluation Project (2022),
       http://www.lnhb.fr/ddep_wg/

[99]   INTERNATIONAL ATOMIC ENERGY AGENCY, Live Chart of Nuclides: nuclear structure and decay data (2022),
       https://www-nds.iaea.org/relnsd/vcharthtml/VChartHTML.html

[100] BROOKHAVEN NATIONAL LABORATORY, National Nuclear Data Center (2022),
       https://www.nndc.bnl.gov/

[101] LABORATOIRE NATIONAL HENRI BECQUEREL FRANCE, Atomic and Nuclear data (2022),
       http://www.lnhb.fr/en/

[102] NUCLEAR INSTITUTE OF STANDARDS AND TECHNOLOGY, P.M.L., Radionuclide Half-Life Measurements Data (2022),
       https://www.nist.gov/pml/radionuclide-half-life-measurements/radionuclide-half-life-measurements-data

[103] PEARCE, A., Recommended Nuclear Decay Data, National Physical Laboratory (NPL) Report IR 6, Teddington, Middlesex (2008).

[104] INTERNATIONAL ORGANIZATION FOR STANDARDIZATION, ISO 8601 Date and time — Representations for information interchange, (2019).

[105] PODGORSAK, M.B., DeWERD, L.A., THOMADSEN, B.R., PALIWAL, B.R., Thermal and scatter effects on the radiation sensitivity of well-type chambers used for high dose rate Ir-192 calibrations, Med. Phys. **19** (1992) 1311–1314.

[106] ATTIX, F.H., Determination of Aion and Pion in the new AAPM radiotherapy dosimetry protocol, Med. Phys. **11** (1984) 714–716.

[107] McEWEN, M., TG-51 Addendum – Overview and Implementation, Med. Phys. **39** (2012) 4006.

[108] INTERNATIONAL ATOMIC ENERGY AGENCY, Dosimetry of Small Static Fields Used in External Beam Radiotherapy. An International Code of Practice for Reference and Relative Dose Determination, Technical Reports Series No. 483, IAEA, Vienna (2017).

[109] McEWEN, M.R., Measurement of ionization chamber absorbed dose k(Q) factors in megavoltage photon beams, Med. Phys. **37** (2010) 2179–2193.

[110] POIRIER, A., DOUYSSET, G., Influence of ambient humidity on the current delivered by air-vented ionization chambers revisited, Phys. Med. Biol. **51** (2006) 4995–5006.

[111] INTERNATIONAL ATOMIC ENERGY AGENCY, Absorbed Dose Determination in External Beam Radiotherapy: An International Code of Practice for Dosimetry Based on Standards of Absorbed Dose to Water - Rev.1, Technical Reports Series No. 398, Vienna (in press).

[112] SMITH, B.R., DeWERD, L.A., CULBERSON, W.S., On the stability of well-type ionization chamber source strength calibration coefficients, Med. Phys. **47** (2020) 4491–4501.

[113] VANDANA, S., SHARMA, S.D., Long term response stability of a well-type ionization chamber used in calibration of high dose rate brachytherapy sources, J. Med. Phys. **35** (2010) 100–103.

[114] MORGAN, A.M., et al., IPEM guidelines on dosimeter systems for use as transfer instruments between the UK primary dosimetry standards laboratory (NPL) and radiotherapy centres, Phys. Med. Biol. **45** (2000) 2445–2457.

[115] DOUYSSET, G., OSTROWSKY, A., DELAUNAY, F., Some unexpected behaviours of PTW/Nucletron well-type ionization chambers, Phys. Med. Biol. **53** (2008) N269–N275.

[116] MUKWADA, G., NEVERI, G., ALKHATIB, Z., WATERHOUSE, D.K., EBERT, M., Commissioning of a well-type chamber for HDR and LDR brachytherapy applications: a review of methodology and outcomes, Australas. Phys. Eng. Sci. Med. **39** (2016) 167–175.

[117] SCHERER, H., DRUNG, D., KRAUSE, C., GOTZ, M., BECKER, U., Electrometer calibration with sub-part-per-million uncertainty, IEEE T. Instrum. Meas. **68** (2019) 1887–1894.

[118] VENSELAAR, J., PÉREZ-CALATAYUD, J., A Practical Guide to Quality Control of Brachytherapy Equipment, ESTRO Booklet No. 8, ESTRO, Brussels (2004).

[119] BOCHUD, F., et al., Dosimetry and Quality Assurance in High Dose Rate Brachytherapy with Iridium-192. Recommendations No.13 of the Swiss Society for Radiobiology and Medical Physics, (2005).

[120] CORMACK, R.A., Quality assurance issues for computed tomography-, ultrasound-, and magnetic resonance imaging-guided brachytherapy, Int. J. Radiat. Oncol. Biol. Phys. **71** (2008) S136–S141.

[121] WILLIAMSON, J.F., Current brachytherapy quality assurance guidance: does it meet the challenges of emerging image-guided technologies?, Int. J. Radiat. Oncol. Biol. Phys. **71** (2008) S18–S22.

[122] INSTITUTE OF PHYSICS AND ENGINEERING IN MEDICINE, PATEL, I., WESTON, S.J., PALMER, A.L., Physics aspects of quality control in radiotherapy (2nd edn), IPEM Rep. 81, IPEM, York (2018).

[123] STEENHUIJSEN, J., et al., Code of Practice for Quality Assurance of Brachytherapy with Ir-192 Afterloaders. Report 30 of the Netherlands Commission on Radiation Dosimetry, (2018).

[124] INTERNATIONAL ORGANIZATION FOR STANDARDIZATION, General Requirements for the Competence of Testing and Calibration Laboratories, ISO/IEC 17025, Geneva (2017).

[125] INTERNATIONAL ATOMIC ENERGY AGENCY, Calibration of Reference Dosimeters for External Beam Radiotherapy, Technical Reports Series No. 469, IAEA, Vienna (2009).

[126] KLEIN, E.E., et al., Task Group 142 report: quality assurance of medical accelerators, Med. Phys. **36** (2009) 4197–4212.

[127] INTERNATIONAL COMMISSION ON RADIATION UNITS AND MEASUREMENTS, International vocabulary of metrology – Basic and general concepts and associated terms (VIM), JCGM 200:2012(E/F), BIPM, (2012).

[128] JUDGE, S., BURNS, D., KESSLER, C., TOROI, P., MSIMANG, Z., The International Measurement System for Radiation Dosimetry, IAEA/WHO SSDL Newsletter. **73** (2021) 5–7.

[129] INTERNATIONAL COMMITTEE FOR WEIGHTS AND MEASURES, Key Comparison Database, BIPM, (1999–2015).

[130] INTERNATIONAL ATOMIC ENERGY AGENCY, SSDL NETWORK CHARTER (2nd edn), IAEA, Vienna (2018).

[131] INTERNATIONAL ATOMIC ENERGY AGENCY, SSDL Network (2021),
https://ssdl.iaea.org/Home/Members

[132] BUREAU INTERNATIONAL DES POIDS ET MESURES, Mutual recognition of national measurement standards and of calibration and measurement certificates issued by national metrology institutes, BIPM, Paris (1999).

[133] BUREAU INTERNATIONAL DES POIDS ET MESURES, Mutual recognition of national measurement standards and of calibration and measurement certificates issued by national metrology institutes – Technical supplement to the arrangement, CIPM, Paris (2003).

[134] ALVAREZ, J.T., et al., Comparison BIPM.RI(I)-K8 of high dose-rate Ir-192 brachytherapy standards for reference air kerma rate of the VSL and the BIPM, Metrologia **51** (2014) 06022.

[135] ALVAREZ, J.T., SANDER, T., DE POOTER, J.A., ALLISY-ROBERTS, P.J., KESSLER, C., Comparison BIPM.RI(I)-K8 of high dose rate Ir-192 brachytherapy standards for reference air kerma rate of the NPL and the BIPM, Metrologia **51** (2014) 06024.

[136] KESSLER, C., ALLISY-ROBERTS, P.J., SELBACH, H.J., Comparison BIPM.RI(I)-K8 of high dose-rate Ir-192 brachytherapy standards for reference air kerma rate of the PTB and the BIPM, Metrologia **52** (2015) 06005.

[137] KESSLER, C., DOWNTON, B., MAINEGRA-HING, E., Comparison BIPM.RI(I)-K8 of high dose-rate Ir-192 brachytherapy standards for reference air kerma rate of the NRC and the BIPM, Metrologia **52** (2015) 06013

[138] KESSLER, C., KUROSAWA, T., MIKAMOTO, T., Comparison BIPM.RI(I)-K8 of high dose-rate Ir-192 brachytherapy standards for reference air kerma rate of the NMIJ and the BIPM, Metrologia **53** (2016) 06001.

[139] BUREAU INTERNATIONAL DES POIDS ET MESURES (BIPM), Key Comparison Database (KCDB) (2021),
https://www.bipm.org/kcdb/

[140] ROZENFELD, M., JETTE, D., Quality assurance of radiation dosage - usefulness of redundancy, Radiology **150** (1984) 241–244.

[141] BIDMEAD, A.M., et al., The IPEM code of practice for determination of the reference air kerma rate for HDR 192Ir brachytherapy sources based on the NPL air kerma standard, Phys. Med. Biol. **55** (2010) 3145–3159.

[142] DeWERD, L.A., et al., A dosimetric uncertainty analysis for photon-emitting brachytherapy sources: report of AAPM Task Group No. 138 and GEC-ESTRO, Med. Phys. **38** (2011) 782–801.

[143] KUTCHER, G.J., et al., Comprehensive QA for radiation oncology: report of AAPM Radiation Therapy Committee Task Group 40, Med. Phys. **21** (1994) 581–618.

[144] SOARES, C.G., DOUYSSET, G., MITCH, M.G., Primary standards and dosimetry protocols for brachytherapy sources, Metrologia **46** (2009) S80–S98.

[145] SCHÜLLER, A., MEIER, M., SELBACH, H.J., ANKERHOLD, U., A radiation quality correction factor k(Q) for well-type ionization chambers for the measurement of the reference air kerma rate of 60Co HDR brachytherapy sources, Med. Phys. **42** (2015) 4285–4294.

[146] CHANG, L., HO, S.Y., CHUI, C.S., DU, Y.C., CHEN, T., Room scatter factor modelling and measurement error analysis of 192Ir HDR calibration by a Farmer chamber, Phys. Med. Biol. **52** (2007) 871–877.

[147] BALTAS, D., et al., Comparison of calibration procedures for Ir-192 high-dose-rate brachytherapy sources, Int. J. Radiat. Oncol. Biol. Phys. **43** (1999) 653–661.

[148] GRIFFIN, S.L., DeWERD, L.A., MICKA, J.A., BOHM, T.D., The effect of ambient pressure on well-type chamber response: experimental results with empirical correction factors, Med. Phys. **32** (2005) 700–709.

[149] BOHM, T.D., GRIFFIN, S.L., DELUCA, P.M., Jr., DeWERD, L.A., The effect of ambient pressure on well-type chamber response: Monte Carlo calculated results for the HDR 1000 plus, Med. Phys. **32** (2005) 1103–1114.

[150] TORNERO-LOPEZ, A.M., et al., Dependence with air density of the response of the PTW SourceCheck ionization chamber for low energy brachytherapy sources, Med. Phys. **40** (2013) 122103.

[151] TORRES DEL RIO, J., TORNERO-LOPEZ, A.M., GUIRADO, D., PEREZ-CALATAYUD, J., LALLENA, A.M., Air density dependence of the response of the PTW SourceCheck 4pi ionization chamber for ^{125}I brachytherapy seeds, Phys. Med. **38** (2017) 93–97.

[152] WATT, E., SPENCER, D.P., MEYER, T., Technical Note: Empirical altitude correction factors for well-type chamber measurements of permanent prostate and breast seed implant sources, Med. Phys. **44** (2017) 5517–5521.

[153] FORASTERO RODRIGUEZ, C., et al., Air density dependence of the response of the PTW sourcecheck 4PI ionization chamber to Pd-103 brachytherapy sources, Phys. Med. **52** (2018) 142.

[154] LAMBECK, J., KENNAN, W., DeWERD, L.A., Effect of well-type chamber altitude pressure corrections for cesium Blu (131) Cs and CivaDot (103) Pd brachytherapy sources, Med. Phys. **48** (2021) 5584–5592.

[155] LA RUSSA, D.J., McEWEN, M., ROGERS, D.W., An experimental and computational investigation of the standard temperature-pressure correction factor for ion chambers in kilovoltage x rays, Med. Phys. **34** (2007) 4690–4699.

[156] ALVAREZ ROMERO, J.T., DE LA CRUZ HERNANDEZ, D., CABRERA VERTTI, R., The dependence of NKR versus KR: the initial, thermal, volumetric recombination and screening effect on the efficiency of collected charges on the calibration of si HDR1000 plus well-type chambers with ^{192}Ir HDR sources, Biomed. Phys. Eng. Express. **8** (2022) 027002.

[157] SHIPLEY, D.R., SANDER, T., NUTBROWN, R.F., Source geometry factors for HDR 192Ir brachytherapy secondary standard well-type ionization chamber calibrations, Phys. Med. Biol. **60** (2015) 2573–2586.

[158] RILEY, A.D., PIKE, T.L., MICKA, J.A., FULKERSON, R.K., DeWERD, L.A., Determination of air-kerma strength for the Ir-192 GammaMedplus iX pulsed-dose-rate brachytherapy source, Med. Phys. **40** (2013) 071732.

[159] RASMUSSEN, B.E., DAVIS, S.D., SCHMIDT, C.R., MICKA, J.A., DeWERD, L.A., Comparison of air-kerma strength determinations for HDR ^{192}Ir sources, Med. Phys. **38** (2011) 6721–6729.

[160] HACKETT, S.L., DAVIS, B., NIXON, A., WYATT, R., Constancy checks of well-type ionization chambers with external-beam radiation units, J. Appl. Clin. Medical. Phys. **16** (2015) 508-514.

[161] BALTAS, D., SAKELLIOU, L., ZAMBOGLOU, N., The Physics of Modern Brachytherapy for Oncology, CRC Press, Boca Raton (2006).

[162] DIEZ, P., et al., A multicentre audit of HDR/PDR brachytherapy absolute dosimetry in association with the INTERLACE trial (NCT015662405), Phys. Med. Biol. **62** (2017) 8832–8849.

[163] DEMPSEY, C., et al., ACPSEM brachytherapy working group recommendations for quality assurance in brachytherapy, Australas. Phys. Eng. Sci. Med. **36** (2013) 387–396.

[164] SELTZER, S.M., et al., New national air-kerma-strength standards for ^{125}I and ^{103}Pd brachytherapy seeds, J. Res. Natl. Inst. Stand. Technol. **108** (2003) 337–358.

[165] SELTZER, S.M., et al., Erratum: New national air-kerma-strength standards for ^{125}I and ^{103}Pd brachytherapy seeds, J. Res. Natl. Inst. Stand. Technol. **109** (2004) 301.

[166] CULBERSON, W.S., DeWERD, L.A., ANDERSON, D.R., MICKA, J.A., Large-volume ionization chamber with variable apertures for air-kerma measurements of low-energy radiation sources, Rev. Sci. Instrum. **77** (2006) 015105.

[167] INTERNATIONAL COMMITTEE FOR WEIGHTS AND MEASURES, Evaluation of Measurement Data - Guide to the Expression of Uncertainty in Measurement, JCGM 100:2008, BIPM, Paris (2008).

[168] INTERNATIONAL ATOMIC ENERGY AGENCY, Measurement Uncertainty, IAEA-TECDOC-1585, Vienna (2008).

[169] DEUTSCHES INSTITUT FÜR NORMUNG (DIN), Dosimetry for Photon Brachytherapy - Part 2: Radiation sources, source calibration, source test and dose calculation, DIN 6803–2, Berlin (2020).

[170] PASTOR-SANCHIS, V., et al., Experimental validation of the Valencia-type applicators developed for the BEBIG HDR afterloader Saginova, Med. Phys. **44** (2017) 3176.

[171] GRANERO, D., et al., Dosimetric relevance of the Valencia and Leipzig HDR applicators plastic cap, Med. Phys. **43** (2016) 3476.

[172] GRANERO, D., et al., Commissioning and quality assurance procedures for the HDR Valencia skin applicators, J. Contemp. Brachytherapy **8** (2016) 441–447.

[173] ARYAL, P., CHEN, S., AGARWAL, M., ZHOU, J., LASIO, G., Commissioning of HDR Valencia applicators, Med. Phys. **46** (2019) e549.

[174] PEREZ-CALATAYUD, J., et al., A dosimetric study of Leipzig applicators, Int. J. Radiat. Oncol. Biol. Phys. **62** (2005) 579–584.

[175] NIU, H., HSI, W., CHU, J., KIRK, M., KOUWENHOVEN, E., Dosimetric characteristics of the HDR Leipzig applicator in surface radiation treatments, Med. Phys. 31 (2004) 1771–1772.

[176] HSI, W., NIU, H., CHU, J., 3D analytical dosimetric model for radiation treatment planning with the HDR Leipzig surface applicator, Med. Phys. 31 (2004) 1808.

[177] PEREZ-CALATAYUD, J., et al., Design and evaluation of an HDR skin applicator with flattening filter, Radiother. Oncol. 84 (2007) S145.

[178] OUHIB, Z., et al., Aspects of dosimetry and clinical practice of skin brachytherapy: The American Brachytherapy Society working group report, Brachytherapy 14 (2015) 840–858.

[179] NACHMAN, J., et al., Disparate histologic responses in simultaneously resected primary and metastatic osteosarcoma following intravenous neoadjuvant chemotherapy, J. Clin. Oncol. 5 (1987) 1185–1190.

[180] FULKERSON, R.K., MICKA, J.A., DeWERD, L.A., Dosimetric characterization and output verification for conical brachytherapy surface applicators. Part II. High dose rate 192Ir sources, Med. Phys. 41 (2014) 022104.

[181] FULKERSON, R.K., et al., Surface brachytherapy: Joint report of the AAPM and the GEC-ESTRO Task Group No. 253, Med. Phys. 47 (2020) e951–e987.

[182] DeWERD, L.A., CULBERSON, W.S., MICKA, J.A., SIMIELE, S.J., A modified dose calculation formalism for electronic brachytherapy sources, Brachytherapy 14 (2015) 405-408.

[183] EURAMET, Primary Standards and Traceable Measurement Methods for X-ray Emitting Electronic Brachytherapy Devices, (2019-2022), Publishable Summary available at http://www.ebt-empir.eu/, 25 November 2020

[184] HIATT, J.R., RIVARD, M.J., HUGHES, H.G., Simulation evaluation of NIST air-kerma rate calibration standard for electronic brachytherapy, Med. Phys. 43 (2016) 1119–1129.

[185] DELANEY, T.F., et al., Intraoperative dural irradiation by customized (192)iridium and (90)yttrium brachytherapy plaques, Int. J. Radiat. Oncol. Biol. Phys. 57 (2003) 239-245.

[186] ZUCKERMAN, S.L., LIM, J., YAMADA, Y., BILSKY, M.H., LAUFER, I., Brachytherapy in spinal tumors: a systematic review, World Neurosurg. 118 (2018) E235–E244.

[187] NATH, R., et al., Intravascular brachytherapy physics: Report of the AAPM Radiation Therapy Committee Task Group No. 60, Med. Phys. 26 (1999) 119–152.

[188] ASTRAHAN, M.A., A patch source model for treatment planning of ruthenium ophthalmic applicators, Med. Phys. 30 (2003) 1219–1228.

[189] WILLIAMSON, J., COURSEY, B.M., DeWERD, L.A., HANSON, W.F., NATH, R., Dosimetric prerequisites for routine clinical use of new low energy photon interstitial brachytherapy sources. Recommendations of the American Association of Physicists in Medicine Radiation Therapy Committee. Ad Hoc Subcommittee of the Radiation Therapy Committee, Med. Phys. 25 (1998) 2269–2270.

[190] DeWERD, L.A., et al., Procedures for establishing and maintaining consistent air-kerma strength standards for low-energy, photon-emitting brachytherapy sources: recommendations of the Calibration Laboratory Accreditation Subcommittee of the American Association of Physicists in Medicine, Med. Phys. 31 (2004) 675–681.

[191] SELTZER, S.M., BERGSTROM, P.M., Jr., Changes in the U.S. Primary Standards for the Air kerma from gamma-ray beams, J. Res. Natl. Inst. Stand. Technol. **108** (2003) 359–381.

[192] SHEN, H., CULBERSON, W.S., ROSS, C.K., Technical Note: An investigation of polarity effects for wide-angle free-air chambers, Med. Phys. **43** (2016) 4106–4112.

[193] SELBACH, H.J., KRAMER, H.M., CULBERSON, W.S., Realization of reference air-kerma rate for low-energy photon sources, Metrologia **45** (2008) 422–428.

[194] BOHM, J., SCHNEIDER, U., Review of extrapolation chamber measurements of beta-rays and low-energy X-rays, Radiat. Prot. Dosimetry **14** (1986) 193–198.

[195] AUBINEAU-LANIECE, I., et al., LNE-LNHB air-kerma and absorbed dose to water primary standards for low dose-rate I-125 brachytherapy sources, Metrologia **49** (2012) S189–S192.

[196] ROSSITER, M.J., WILLIAMS, T.T., BASS, G.A., Air kerma rate calibration of small sources of Co-60, Cs-137, Ra-226 and Ir-192, Phys. Med. Biol. **36** (1991) 279–284.

[197] SEPHTON, J.P., et al., Calibration of the NPL secondary standard radionuclide calibrator for Ir-192 brachytherapy sources, Phys. Med. Biol. **38** (1993) 1157–1164.

[198] INTERNATIONAL ORGANIZATION FOR STANDARDIZATION, X and gamma reference radiations for calibrating dosemeters and dose ratemeters and for determining their response as a function of photon energy, Part 1: characteristics of the radiations and their methods of production, ISO 4037, Geneva (1993).

[199] VERHAEGEN, F., VANDIJK, E., THIERENS, H., AALBERS, A., SEUNTJENS, J., Calibration of low activity Ir-192 brachytherapy sources in terms of reference air kerma rate with large volume spherical ionization chambers, Phys. Med. Biol. **37** (1992) 2071–2082.

[200] PIERMATTEI, A., AZARIO, L., Applications of the Italian protocol for the calibration of brachytherapy sources, Phys. Med. Biol. **42** (1997) 1661–1669.

[201] OBORIN, A.V., TROFIMCHUK, S.G., VILLEVALDE, A.Y., YAKOVENKO, A.A., Air kerma rate measurement for I-125 medical microsources, Meditsinskaya Fizika. **71** (2016) 40–48.

[202] VILLEVALDE, A.Y., OBORIN, A.V., TROFIMCHUK, S.G., Metrological support of dosimetry measurements in brachytherapy, Ukrainian Metrological Journal, (2017).

[203] GOETSCH, S.J., ATTIX, F.H., PEARSON, D.W., THOMADSEN, B.R., Calibration of Ir-192 high-dose-rate afterloading systems, Med. Phys. **18** (1991) 462–467.

[204] STUMP, K.E., DeWERD, L.A., MICKA, J.A., ANDERSON, D.R., Calibration of new high dose rate 192Ir sources, Med. Phys. **29** (2002) 1483–1488.

[205] MAINEGRA-HING, E., ROGERS, D.W., On the accuracy of techniques for obtaining the calibration coefficient NK of ^{192}Ir HDR brachytherapy sources, Med. Phys. **33** (2006) 3340–3347.

[206] KUMAR, S., SRINIVASAN, P., SHARMA, S.D., SUBBAIAH, K.V., MAYYA, Y.S., Evaluation of scatter contribution and distance error by iterative methods for strength determination of HDR Ir-192 brachytherapy source, Med. Dosim. **35** (2010) 230–237.

[207] KUMAR, S., SRINIVASAN, P., SHARMA, S.D., MAYYA, Y.S., A simplified analytical approach to estimate the parameters required for strength determination of HDR Ir-192 brachytherapy sources using a Farmer-type ionization chamber, Appl. Radiat. Isotopes. **70** (2012) 282–289.

[208] SELBACH, H.J., New calibration device for ^{192}Ir and ^{60}Co-brachytherapy radiation sources [In German]. Tagungsband der 37. Jahrestagung der Deutschen Gesellschaft für Medizinische Physik e.V, (2006).

[209] BÜERMANN, L., KRAMER, H.M., SCHRADER, H., SELBACH, H.J., Activity determination of ^{192}Ir solid sources by ionization chamber measurements using calculated corrections of self-absorption, Nucl. Instrum. Methods. Phys. Res. A. **339** (1994) 369–376.

[210] MARECHAL, M.H., FERREIRA, I.H., PEIXOTO, J.G., SIBATA, C.H., DE ALMEIDA, C.E., A method to determine the air kerma calibration factor for thimble ionization chambers used for Ir-192 HDR source calibration, Phys. Medica. **19** (2003) 131–135.

[211] DI PRINZIO, R., DE ALMEIDA, C.E., Air kerma standard for calibration of well-type chambers in Brazil using Ir-192 HDR sources and its traceability, Med. Phys. **36** (2009) 953–960.

[212] DOUYSSET, G., et al., Comparison of dosimetric standards of USA and France for HDR brachytherapy, Phys. Med. Biol. **50** (2005) 1961–1978.

[213] DOUYSSET, G., GOURIOU, J., DELAUNAY, F., Dose metrology for high dose rate brachytherapy: from the definition of the national standard towards transfer to users [in French], Revue Francaise de Metrologie **2** (2007) 3–10.

[214] VAN DIJK, E., KOLKMAN-DEURLOO, I.K.K., DAMEN, P.M.G., Determination of the reference air kerma rate for Ir-192 brachytherapy sources and the related uncertainty, Med. Phys. **31** (2004) 2826–2833.

[215] PETERSEN, J.J., VAN DIJK, E., AALBERS, A., Comparison of methods for derivation of ^{192}Ir calibration factors for the NE 2561 & 2571 ionisation chambers, Report S-EI-94.01, Utrecht, The Netherlands: NMi Van Swinden Laboratorium (1994).

[216] PODER, J., et al., High dose rate brachytherapy source measurement intercomparison, Australas. Phys. Eng. Sci. Med. **40** (2017) 377–383.

[217] BUTLER, D., HAWORTH, A., SANDER, T., TODD, S., Comparison of Ir-192 air kerma calibration coefficients derived at ARPANSA using the interpolation method and at the National Physical Laboratory using a direct measurement, Australas. Phys. Eng. Sci. Med. **31** (2008) 332–338.

[218] SANDER, T., NUTBROWN, R.F., The NPL air kerma primary standard TH100C for high dose rate 192Ir brachytherapy sources, NPL Report. DQL-RD 004, National Physical Laboratory, Teddington, UK (2006).

[219] KUMAR, S., SRINIVASAN, P., SHARMA, S.D., Calibration coefficient of reference brachytherapy ionization chamber using analytical and Monte Carlo methods, Appl. Radiat. Isot. **68** (2010) 1108–1115.

[220] CHU, W.H., YUAN, M.C., LEE, J.H., LIN, Y.C., Reference air kerma rate calibration system for high dose rate Ir-192 brachytherapy sources in Taiwan, Rad. Phys. Chem. **140** (2017) 361–364.

[221] KIM, Y., YI, C.Y., KIM, I.J., SEONG, Y.M., Changes of KRISS primary standards by implementing ICRU 90 recommendation, J. Korean Phys. Soc. **78** (2021) 842–848.

[222] SCHNEIDER, T., SELBACH, H.J., Realisation of the absorbed dose to water for I-125 interstitial brachytherapy sources, Radiother. Oncol. **100** (2011) 442–445.

[223] SCHNEIDER, T., A method to determine the water kerma in a phantom for x-rays with energies up to 40 keV, Metrologia **46** (2009) 95–100.

[224] SCHNEIDER, T., A robust method for determining the absorbed dose to water in a phantom for low-energy photon radiation, Phys. Med. Biol. **56** (2011) 3387–3402.

[225] TONI, M.P., et al., Direct determination of the absorbed dose to water from I-125 low dose-rate brachytherapy seeds using the new absorbed dose primary standard developed at ENEA-INMRI, Metrologia **49** (2012) S193–S197.

[226] MALIN, M.J., PALMER, B.R., DeWERD, L.A., Absolute measurement of LDR brachytherapy source emitted power: Instrument design and initial measurements, Med. Phys. **43** (2016) 796–806.

[227] STUMP, K.E., DeWERD, L.A., RUDMAN, D.A., SCHIMA, S.A., Active radiometric calorimeter for absolute calibration of radioactive sources, Rev. Sci. Instrum. **76** (2005) 033504.

[228] KRAUSS, A., The PTB water calorimeter for the absolute determination of absorbed dose to water in Co-60 radiation, Metrologia **43** (2006) 259–272.

[229] BAMBYNEK, M., KRAUSS, A., Determination of absorbed dose to water for ^{192}Ir HDR brachytherapy sources in near-field geometry, ed. PTB report: advanced metrology for cancer therapy, Proc. Int. Conf. Braunschweig, KAPSCH, R.P., PTB-Dos-56. Braunschweig, Germany: PTB (2011).

[230] DE PREZ, L.A., DE POOTER, J.A., Development of the VSL water calorimeter as a primary standard for absorbed dose to water measurements for HDR brachytherapy sources, ed. PTB report: advanced metrology for cancer therapy, Proc. Int. Conf. Braunschweig, KAPSCH, R.P., PTB-Dos-56. Braunschweig, Germany: PTB (2011).

[231] GUERRA, A.S., et al., A standard graphite calorimeter for dosimetry in brachytherapy with high dose rate Ir-192 sources, Metrologia **49** (2012) S179–S183.

[232] SANDER, T., et al., NPL's new absorbed dose standard for the calibration of HDR Ir-192 brachytherapy sources, Metrologia **49** (2012) S184–S188.

[233] SELBACH, H.J., et al., Corrigendum: Experimental determination of the dose rate constant for selected I-125- and Ir-192-brachytherapy sources (2012 Metrologia 49 S219–22), Metrologia **51** (2014) 127.

[234] SELBACH, H.J., et al., Experimental determination of the dose rate constant for selected I-125- and Ir-192-brachytherapy sources, Metrologia **49** (2012) S219–S222.

[235] AUSTERLITZ, C., et al., Determination of absorbed dose in water at the reference point $D(r_0, \theta_0)$ for an Ir-192 HDR brachytherapy source using a Fricke system, Med. Phys. **35** (2008) 5360–5365.

[236] EL GAMAL, I., COJOCARU, C., MAINEGRA-HING, E., McEWEN, M., The Fricke dosimeter as an absorbed dose to water primary standard for Ir-192 brachytherapy, Phys. Med. Biol. **60** (2015) 4481–4495.

[237] FRANCO, L., GAVAZZI, S., COELHO, M., DE ALMEIDA, C.E., Determination of the Fricke G Value for HDR [192]Ir Sources Using Ionometric Measurements, (Standards, applications and quality assurance in medical radiation dosimetry (IDOS) - Proceedings of an International Symposium, Vienna, 9–12 November 2010), Vol. 1, IAEA, 111–119.

[238] McEWEN, M., GAMAL, I., MAINEGRA-HING, E., COJOCARU, C., Determination of the radiation chemical yield (G) for the Fricke chemical dosimetry system in photon and electron beams, Report NRC-PIRS-1980, Ionizing Radiation Standards, National Research Council Canada, Ottawa, Canada (2014).

[239] DE ALMEIDA, C.E., et al., A feasibility study of Fricke dosimetry as an absorbed dose to water standard for Ir-192 HDR sources, Plos One 9 (2014) e115155.

[240] SALATA, C., et al., Validating Fricke dosimetry for the measurement of absorbed dose to water for HDR Ir-192 brachytherapy: a comparison between primary standards of the LCR, Brazil, and the NRC, Canada, Phys. Med. Biol. 63 (2018) 085004.

[241] HANSEN, J.B., CULBERSON, W.S., DeWERD, L.A., Windowless extrapolation chamber measurement of surface dose rate from a Sr-90/Y-90 ophthalmic applicator, Radiat. Meas. 108 (2018) 34–40.

[242] INTERNATIONAL ORGANIZATION FOR STANDARDIZATION, Clinical Dosimetry – Beta radiation sources for brachytherapy, Geneva (2009).

[243] SOARES, C.G., HALPERN, D.G., WANG, C.K., Calibration and characterization of beta-particle sources for intravascular brachytherapy, Med. Phys. 25 (1998) 339–346.

[244] DINSMORE, M., et al., A new miniature x-ray source for interstitial radiosurgery: device description, Med. Phys. 23 (1996) 45–52.

[245] DOUGLAS, R.M., et al., Dosimetric results from a feasibility study of a novel radiosurgical source for irradiation of intracranial metastases, Int. J. Radiat. Oncol. Biol. Phys. 36 (1996) 443–450.

[246] BEATTY, J., et al., A new miniature x-ray device for interstitial radiosurgery: dosimetry, Med. Phys. 23 (1996) 53–62.

[247] ARMOOGUM, K.S., PARRY, J.M., SOULIMAN, S.K., SUTTON, D.G., MACKAY, C.D., Functional intercomparison of intraoperative radiotherapy equipment - Photon Radiosurgery System, Radiat. Oncol. 2 (2007) 11.

[248] SCHNEIDER, T., SELBACH, H.J., ROUIJAA, M., Absolute dosimetry for brachytherapy with the Intrabeam® miniature X-ray devices, Radiother. Oncol. 96 (2010) S573.

[249] EATON, D.J., Quality assurance and independent dosimetry for an intraoperative x-ray device, Med. Phys. 39 (2012) 6908–6920.

[250] SCHNEIDER, F., et al., A novel device for intravaginal electronic brachytherapy, Int. J. Radiat. Oncol. Biol. Phys. 74 (2009) 1298–1305.

[251] WENZ, F., et al., Kypho-IORT - a novel approach of intraoperative radiotherapy during kyphoplasty for vertebral metastases, Radiat. Oncol. 5 (2010) Article number 11.

[252] SCHNEIDER, F., et al., Development of a novel method for intraoperative radiotherapy during kyphoplasty for spinal metastases (Kypho-Iort), Int. J. Radiat. Oncol. Biol. Phys. 81 (2011) 1114–1119.

[253] RIVARD, M.J., DAVIS, S.D., DeWERD, L.A., RUSCH, T.W., AXELROD, S., Calculated and measured brachytherapy dosimetry parameters in water for the Xoft Axxent X-Ray Source: an electronic brachytherapy source, Med. Phys. **33** (2006) 4020–4032.

[254] LIU, D., et al., Spectroscopic characterization of a novel electronic brachytherapy system, Phys. Med. Biol. **53** (2008) 61–75.

[255] MEHTA, V.K., et al., Experience with an electronic brachytherapy technique for intracavitary accelerated partial breast irradiation, Am. J Clin. Oncol. - Cancer Clinical Trials **33** (2010) 327–335.

[256] DICKLER, A., Xoft Axxent® electronic brachytherapy - a new device for delivering brachytherapy to the breast, Nat. Clin. Pract. Oncol. **6** (2009) 138–142.

[257] DICKLER, A., et al., A dosimetric comparison of Xoft Axxent electronic brachytherapy and iridium-192 high-dose-rate brachytherapy in the treatment of endometrial cancer, Brachytherapy **7** (2008) 351–354.

[258] RONG, Y., WELSH, J.S., Surface applicator calibration and commissioning of an electronic brachytherapy system for nonmelanoma skin cancer treatment, Med. Phys. **37** (2010) 5509–5517.

[259] RICHARDSON, S., GARCIA-RAMIREZ, J., LU, W., MYERSON, R.J., PARIKH, P., Design and dosimetric characteristics of a new endocavitary contact radiotherapy system using an electronic brachytherapy source, Med. Phys. **39** (2012) 6838–6846.

[260] CROCE, O., et al., Contact radiotherapy using a 50 kV X-ray system: Evaluation of relative dose distribution with the Monte Carlo code PENELOPE and comparison with measurements, Rad. Phys. Chem. **81** (2012) 609–617.

[261] GERARD, J.P., et al., A brief history of contact X-ray brachytherapy 50 kVp, Cancer Radiother. **24** (2020) 222–225.

[262] IBANEZ-ROSELLO, B., et al., Failure mode and effects analysis of skin electronic brachytherapy using Esteya (R) unit, J Contemp Brachytherapy **8** (2016) 518–524.

[263] GARCIA-MARTINEZ, T., CHAN, J.P., J.P.C., BALLESTER, F., Dosimetric characteristics of a new unit for electronic skin brachytherapy, J. Contemp. Brachytherapy **6** (2014) 45–53.

[264] MA, C.M., et al., AAPM protocol for 40-300 kV x-ray beam dosimetry in radiotherapy and radiobiology, Med. Phys. **28** (2001) 868–893.

[265] CULBERSON, W.S., et al., Dose-rate considerations for the INTRABEAM electronic brachytherapy system: Report from the American association of physicists in medicine task group no. 292, Med. Phys. **47** (2020) e913–e919.

[266] LAMPERTI, P.J., WYCKOFF, H.O., NBS Free-Air Chamber for Measurement of 10 to 60 Kv X Rays, J. Res. Natl. Bur. Stand. C Eng. Instrum. **69C** (1965) 39–47.

[267] ABUDRA'A, A., et al., Dosimetry formalism and calibration procedure for electronic brachytherapy sources in terms of absorbed dose to water, Phys. Med. Biol. **65** (2020) 145006.

[268] WATSON, P.G.F., POPOVIC, M., SEUNTJENS, J., Determination of absorbed dose to water from a miniature kilovoltage x-ray source using a parallel-plate ionization chamber, Phys. Med. Biol. **63** (2018) 015016.

[269] WATSON, P.G.F., BEKERAT, H., PAPACONSTADOPOULOS, P., DAVIS, S., SEUNTJENS, J., An investigation into the INTRABEAM miniature x-ray source dosimetry using ionization chamber and radiochromic film measurements, Med. Phys. **45** (2018) 4274–4286.

[270] SCHNEIDER, T., RADECK, D., ŠOLC, J., Development of a new primary standard for the realization of the absorbed dose to water for electronic brachytherapy X-ray sources, Brachytherapy **15** (2016) S27–S28.

[271] MORRISON, H., MENON, G., SLOBODA, R.S., Radiochromic film calibration for low-energy seed brachytherapy dose measurement, Med. Phys. **41** (2014) 072101.

[272] SMITH, B.R., MICKA, J.A., AIMA, M., DeWERD, L.A., CULBERSON, W.S., Air-kerma strength determination of an HDR ^{192}Ir source including a geometric sensitivity study of the seven-distance method, Med. Phys. **44** (2017) 311–320.

[273] HEILEMANN, G., NESVACIL, N., BLAICKNER, M., KOSTIUKHINA, N., GEORG, D., Multidimensional dosimetry of ^{106}Ru eye plaques using EBT3 films and its impact on treatment planning, Med. Phys. **42** (2015) 5798–5808.

[274] HERMIDA-LOPEZ, M., BRUALLA, L., Absorbed dose distributions from ophthalmic ^{106}Ru/^{106}Rh plaques measured in water with radiochromic film, Med. Phys. **45** (2018) 1699–1707.

[275] PALMER, A.L., BRADLEY, D.A., NISBET, A., Dosimetric audit in brachytherapy, Br. J. Radiol. **87** (2014) 20140105.

[276] PALMER, A.L., LEE, C., RATCLIFFE, A.J., BRADLEY, D., NISBET, A., Design and implementation of a film dosimetry audit tool for comparison of planned and delivered dose distributions in high dose rate (HDR) brachytherapy, Phys. Med. Biol. **58** (2013) 6623–6640.

[277] CHIUTSAO, S.T., ANDERSON, L.L., Thermoluminescent dosimetry for Pd-103 seeds (model 200) in solid water phantom, Med Phys. **18** (1991) 449–452.

[278] KIROV, A., WILLIAMSON, J.F., MEIGOONI, A.S., ZHU, Y., TLD, diode and Monte Carlo dosimetry of an 192Ir source for high dose-rate brachytherapy, Phys. Med. Biol. **40** (1995) 2015–2036.

[279] KARAISKOS, P., et al., Monte Carlo and TLD dosimetry of an ^{192}Ir high dose-rate brachytherapy source, Med. Phys. **25** (1998) 1975–1984.

[280] MEIGOONI, A.S., SOWARDS, K., SOLDANO, M., Dosimetric characteristics of the InterSource(103) palladium brachytherapy source, Med. Phys. **27** (2000) 1093–1100.

[281] DeWERD, L.A., LIANG, Q., REED, J.L., CULBERSON, W.S., The use of TLDs for brachytherapy dosimetry, Radiat. Meas. **71** (2014) 276–281.

[282] PIESSENS, M., REYNAERT, N., Verification of absolute dose rates for intravascular brachytherapy beta sources, Phys. Med. Biol. **45** (2000) 2219–2231.

[283] REFT, C.S., KUCHNIR, F.T., ROSENBERG, I., MYRIANTHOPOULOS, L.C., Dosimetry of Sr-90 ophthalmic applicators, Med. Phys. **17** (1990) 641–646.

[284] BINDER, W., CHIARI, A., AIGINGER, H., Determination of the dose distribution of an ophthalmic Ru-106 irradiator with TLDs and an eye phantom, Radiat. Prot. Dosimetry **34** (1990) 275–278.

[285] SIDDLE, D., LANGMACK, K., Calibration of strontium-90 eye applicator using a strontium external beam standard, Phys. Med. Biol. **44** (1999) 1597–1608.

[286] KAPP, K.S., STUECKLSCHWEIGER, G.F., KAPP, D.S., HACKL, A.G., Dosimetry of intracavitary placements for uterine and cervical-carcinoma - results of orthogonal film, TLD, and CT-assisted techniques, Radiother. Oncol. **24** (1992) 137–146.

[287] TOYE, W., et al., An in vivo investigative protocol for HDR prostate brachytherapy using urethral and rectal thermoluminescence dosimetry, Radiother. Oncol. **91** (2009) 243–248.

[288] RAFFI, J.A., et al., Determination of exit skin dose for Ir-192 intracavitary accelerated partial breast irradiation with thermoluminescent dosimeters, Med. Phys. **37** (2010) 2693–2702.

[289] ROUE, A., VENSELAAR, J.L.M., FERREIRA, I.H., BRIDIER, A., VAN DAM, J., Developments of a TLD mailed system for remote dosimetry audit for Ir-192 HDR and PDR sources, Radiother. Oncol. **83** (2007) 86–93.

[290] PALMER, A.L., BRADLEY, D.A., NISBET, A., Improving quality assurance of HDR brachytherapy: verifying agreement between planned and delivered dose distributions using DICOM RTDose and advanced film dosimetry, Med. Phys. **41** (2014) 270.

[291] NUNN, A.A., DAVIS, S.D., MICKA, J.A., DeWERD, L.A., LiF : Mg,Ti TLD response as a function of photon energy for moderately filtered x-ray spectra in the range of 20-250 kVp relative to Co-60, Med. Phys. **35** (2008) 1859–1869.

[292] DAVIS, S.D., et al., The response of LiF thermoluminescence dosemeters to photon beams in the energy range from 30 kV X rays to Co-60 gamma rays, Radiat. Prot. Dosimetry **106** (2003) 33–43.

[293] TEDGREN, A.C., HEDMAN, A., GRINDBORG, J.E., CARLSSON, G.A., Response of LiF:Mg,Ti thermoluminescent dosimeters at photon energies relevant to the dosimetry of brachytherapy (< 1 MeV), Med. Phys. **38** (2011) 5539–5550.

[294] TEDGREN, A.C., ELIA, R., HEDTJARN, H., OLSSON, S., CARLSSON, G.A., Determination of absorbed dose to water around a clinical HDR Ir-192 source using LiF:Mg,Ti TLDs demonstrates an LET dependence of detector response, Med. Phys. **39** (2012) 1133–1140.

[295] RODRIGUEZ, M., ROGERS, D.W.O., Effect of improved TLD dosimetry on the determination of dose rate constants for I-125 and Pd-103 brachytherapy seeds, Med. Phys. **41** (2014) 114301.

[296] HAWORTH, A., et al., Comparison of TLD calibration methods for Ir-192 dosimetry, J. Appl. Clin. Medical Phys. **14** (2013) 258–272.

[297] RUSTGI, S.N., Application of a diamond detector to brachytherapy dosimetry, Phys. Med. Biol. **43** (1998) 2085–2094.

[298] NAKANO, T., et al., High dose-rate brachytherapy source localization: positional resolution using a diamond detector, Phys. Med. Biol. **48** (2003) 2133–2146.

[299] LAMBERT, J., et al., In vivo dosimeters for HDR brachytherapy: A comparison of a diamond detector, MOSFET, TLD, and scintillation detector, Med. Phys. **34** (2007) 1759–1765.

[300] ROSSI, G., et al., Monte Carlo and experimental high dose rate Ir-192 brachytherapy dosimetry with microdiamond detectors, Z. Med. Phys. **29** (2019) 272–281.

[301] ROSSI, G., GAINEY, M., KOLLEFRATH, M., HOFMANN, E., BALTAS, D., Suitability of the microDiamond detector for experimental determination of the anisotropy function of high dose rate 192Ir brachytherapy sources, Med. Phys. **47** (2020) 5838–5851.

[302] KAVECKYTE, V., MALUSEK, A., BENMAKHLOUF, H., CARLSSON, G.A., TEDGREN, A.C., Suitability of microDiamond detectors for the determination of absorbed dose to water around high-dose-rate Ir-192 brachytherapy sources, Med. Phys. **45** (2018) 429–437.

[303] KAMPFER, S., CHO, N., COMBS, S.E., WILKENS, J.J., Dosimetric characterization of a single crystal diamond detector in X-ray beams for preclinical research, Z. Med. Phys. **28** (2018) 303–309.

[304] KAVECKYTE, V., et al., Investigation of a synthetic diamond detector response in kilovoltage photon beams, Med. Phys. **47** (2020) 1268–1279.

[305] GARCIA YIP, F., et al., Characterization of small active detectors for electronic brachytherapy dosimetry, J. Instrum. **17** (2022) P03001.

[306] SAINI, A.S., ZHU, T.C., Energy dependence of commercially available diode detectors for in-vivo dosimetry, Med. Phys. **34** (2007) 1704–1711.

[307] ARBER, J.M., SHARPE, P.H.G., Fading characteristics of irradiated alanine pellets - the importance of preirradiation conditioning, Appl. Radiat. Isot. **44** (1993) 19–22.

[308] SHARPE, P.H.G., RAJENDRAN, K., SEPHTON, J.P., Progress towards an alanine/ESR therapy level reference dosimetry service at NPL, Appl. Radiat. Isot. **47** (1996) 1171–1175.

[309] ANTON, M., Uncertainties in alanine/ESR dosimetry at the Physikalisch-Technische Bundesanstalt, Phys. Med. Biol. **51** (2006) 5419–5440.

[310] ZENG, G.G., MCCAFFREY, J.P., The response of alanine to a 150 keV X-ray beam, Rad. Phys. Chem. **72** (2005) 537–540.

[311] WALDELAND, E., HOLE, E.O., SAGSTUEN, E., MALINEN, E., The energy dependence of lithium formate and alanine EPR dosimeters for medium energy x rays, Med. Phys. **37** (2010) 3569–3575.

[312] WALDELAND, E., MALINEN, E., Review of the dose-to-water energy dependence of alanine and lithium formate EPR dosimeters and LiF TL-dosimeters - Comparison with Monte Carlo simulations, Radiat. Meas. **46** (2011) 945–951.

[313] ANTON, M., BUERMANN, L., Relative response of the alanine dosimeter to medium energy x-rays, Phys. Med. Biol. **60** (2015) 6113–6129.

[314] SCHAEKEN, B., CUYPERS, R., GOOSSENS, J., VAN DEN WEYNGAERT, D., VERELLEN, D., Experimental determination of the energy response of alanine pellets in the high dose rate 192Ir spectrum, Phys. Med. Biol. **56** (2011) 6625–6634.

[315] ANTON, M., HACKEL, T., ZINK, K., VON VOIGTS-RHETZ, P., SELBACH, H.J., Response of the alanine/ESR dosimeter to radiation from an Ir-192 HDR brachytherapy source, Phys. Med. Biol. **60** (2015) 175–193.

[316] WAGNER, D., HERMANN, M., HILLE, A., In vivo dosimetry with alanine/electron spin resonance dosimetry to evaluate the urethra dose during high-dose-rate brachytherapy, Brachytherapy **16** (2017) 815–821.

[317] OLSSON, S., BERGSTRAND, E.S., CARLSSON, A.K., HOLE, E.O., LUND, E., Radiation dose measurements with alanine/agarose gel and thin alanine films around a 192Ir brachytherapy source, using ESR spectroscopy, Phys. Med. Biol. **47** (2002) 1333–1356.

[318] TIEN, C.J., EBELING, R., III, HIATT, J.R., CURRAN, B., STERNICK, E., Optically stimulated luminescent dosimetry for high dose rate brachytherapy, Front Oncol. **2** (2012) 91.

[319] SHARMA, R., JURSINIC, P.A., In vivo measurements for high dose rate brachytherapy with optically stimulated luminescent dosimeters, Med. Phys. **40** (2013) 071730.

[320] CASEY, K.E., et al., Development and implementation of a remote audit tool for high dose rate (HDR) Ir-192 brachytherapy using optically stimulated luminescence dosimetry, Med. Phys. **40** (2013) 112102.

[321] MIZUNO, H., et al., Application of a radiophotoluminescent glass dosimeter to nonreference condition dosimetry in the postal dose audit system, Med. Phys. **41** (2014) 112104.

[322] MIZUNO, H., et al., Feasibility study of glass dosimeter postal dosimetry audit of high-energy radiotherapy photon beams, Radiother. Oncol. **86** (2008) 258–263.

[323] WESOLOWSKA, P.E., et al., Characterization of three solid state dosimetry systems for use in high energy photon dosimetry audits in radiotherapy, Radiat. Meas. **106** (2017) 556–562.

[324] RAH, J.E., et al., A comparison of the dosimetric characteristics of a glass rod dosimeter and a thermoluminescent dosimeter for mailed dosimeter, Radiat. Meas. **44** (2009) 18–22.

[325] IZEWSKA, J., BOKULIC, T., KAZANTSEV, P., WESOLOWSKA, P., VAN DER MERWE, D., 50 Years of the IAEA/WHO postal dose audit programme for radiotherapy: what can we learn from 13756 results?, Acta Oncol. **59** (2020) 495–502.

[326] HASHIMOTO, S., NAKAJIMA, Y., KADOYA, N., ABE, K., KARASAWA, K., Energy dependence of a radiophotoluminescent glass dosimeter for HDR Ir-192 brachytherapy source, Med. Phys. **46** (2019) 964–972.

[327] HSU, S.M., et al., Clinical application of radiophotoluminescent glass dosimeter for dose verification of prostate HDR procedure, Med. Phys. **35** (2008) 5558–5564.

[328] NOSE, T., et al., In vivo dosimetry of high-dose-rate brachytherapy: Study on 61 head-and-neck cancer patients using radiophotoluminescence glass dosimeter, Int. J. Radiat. Oncol. Biol. Phys. **61** (2005) 945–953.

[329] KRON, T., METCALFE, P., POPE, J.M., Investigation of the tissue equivalence of gels used for NMR dosimetry, Phys. Med. Biol. **38** (1993) 139–150.

[330] PANTELIS, E., et al., Polymer gel water equivalence and relative energy response with emphasis on low photon energy dosimetry in brachytherapy, Phys. Med. Biol. **49** (2004) 3495–3514.

[331] SELLAKUMAR, P., SAMUEL, E.J.J., SUPE, S.S., Water equivalence of polymer gel dosimeters, Rad. Phys. Chem. **76** (2007) 1108–1115.

[332] BALDOCK, C., et al., Polymer gel dosimetry, Phys. Med. Biol. **55** (2010) R1–R63.

[333] FARHOOD, B., GERAILY, G., ABTAHI, S.M.M., A systematic review of clinical applications of polymer gel dosimeters in radiotherapy, Appl. Radiat. Isot. **143** (2019) 47–59.

[334] MACDOUGALL, N., PITCHFORD, W.G., SMITH, M.A., A systematic review of the precision and accuracy of dose measurements in photon radiotherapy using polymer and Fricke MRI gel dosimetry, Phys. Med. Biol. **47** (2002) R107–R121.

[335] SCHREINER, L.J., Review of Fricke gel dosimeters, J. Phys. Conf. Ser. **3** (2004) 9.

[336] SCHREINER, L.J., True 3D chemical dosimetry (gels, plastics): Development and clinical role, J. Phys. Conf. Ser. **573** (2015) 012003.

[337] WATANABE, Y., et al., Dose distribution verification in high-dose-rate brachytherapy using a highly sensitive normoxic N-vinylpyrrolidone polymer gel dosimeter, Phys. Med. **57** (2019) 72–79.

[338] TACHIBANA, H., et al., End-to-end delivery quality assurance of computed tomography-based high-dose-rate brachytherapy using a gel dosimeter, Brachytherapy **19** (2020) 362–371.

[339] SENKESEN, O., TEZCANLI, E., BUYUKSARAC, B., OZBAY, I., Comparison of 3D dose distributions for HDR Ir-192 brachytherapy sources with normoxic polymer gel dosimetry and treatment planning system, Med. Dosim. **39** (2014) 266–271.

[340] PAPPAS, E., KARAISKOS, P., ZOURARI, K., PEPPA, V., PAPAGIANNIS, P., An experimental commissioning test of brachytherapy MBDCA dosimetry, based on a commercial radiochromic Gel/optical CT system, Med. Phys. **42** (2015) 3536.

[341] OE, A., NEMOTO, M., MIYAZAWA, M., SAHADE, D.A., HAMADA, T., Spatial dose distribution analysis of Co-60 HDR brachytherapy of cervical cancer using an AQUAJOINT (R)-based VIPET polymer gel dosimeter, J. Phys. Conf. Ser. **1305** (2019) 012052.

[342] FAZLI, Z., SADEGHI, M., ZAHMATKESH, M.H., MAHDAVI, S.R., TENREIRO, C., Dosimetric comparison between three dimensional treatment planning system, Monte Carlo simulation and gel dosimetry in nasopharynx phantom for high dose rate brachytherapy, J. Cancer Res. Ther. **9** (2013) 402–409.

[343] EGUCHI, K., et al., A verification of high-dose-rate brachytherapy dose distributions for prostate cancer with a VIPET polymer gel dosimeter, Med. Phys. **45** (2018) e238.

[344] ADINEHVAND, K., RAHATABAD, F.N., Monte-Carlo based assessment of MAGIC, MAGICAUG, PAGATUG and PAGATAUG polymer gel dosimeters for ovaries and uterus organ dosimetry in brachytherapy, nuclear medicine and Tele-therapy, Comput. Meth. Prog. Bio. **159** (2018) 37–50.

[345] PAPPAS, E.P., PEPPA, V., HOURDAKIS, C.J., KARAISKOS, P., PAPAGIANNIS, P., On the use of a novel ferrous Xylenol-orange gelatin dosimeter for HDR brachytherapy commissioning and quality assurance testing, Phys. Med. **45** (2018) 162–169.

[346] NASR, A.T., SCHREINER, L.J., MCAULEY, K.B., Mathematical modeling of the response of polymer gel dosimeters to HDR and LDR brachytherapy radiation, Macromol. Theor. Simul. **21** (2012) 36–51.

[347] SATO, R., DE ALMEIDA, A., MOREIRA, M.V., Cs-137 source dose distribution using the Fricke Xylenol gel dosimetry, Nucl. Instrum. Methods Phys. Res. B. **267** (2009) 842–845.

[348] AZMA, Z., JABERI, R., AGHAMIRI, M.R., ZAHMATKESH, M.H., Investigation of ruthenium-106 beta emitter eye plaques with X-ray CT PAGAT gel dosimetry, Radiother. Oncol. **91** (2009) S43.

[349] AMIN, M.N., et al., A comparison of polyacrylamide gels and radiochromic film for source measurements in intravascular brachytherapy, Br. J. Radiol. 76 (2003) 824–831.

[350] DEBNATH, S.B.C., et al., High resolution small-scale inorganic scintillator detector: HDR brachytherapy application, Med. Phys. **48** (2021) 1485–1496.

[351] KIROV, A.S., et al., Towards two-dimensional brachytherapy dosimetry using plastic scintillator: New highly efficient water equivalent plastic scintillator materials, Med. Phys. **26** (1999) 1515–1523.

[352] KERTZSCHER, G., BEDDAR, S., Inorganic scintillation detectors for Ir-192 brachytherapy, Phys. Med. Biol. **64** (2019) 225018.

[353] LAMBERT, J., MCKENZIE, D.R., LAW, S., ELSEY, J., SUCHOWERSKA, N., A plastic scintillation dosimeter for high dose rate brachytherapy, Phys. Med. Biol. **51** (2006) 5505–5516.

[354] ROSALES, H.M.L., ARCHAMBAULT, L., BEDDAR, S., BEAULIEU, L., Dosimetric performance of a multipoint plastic scintillator dosimeter as a tool for real-time source tracking in high dose rate ^{192}Ir brachytherapy, Med. Phys. **47** (2020) 4477–4490.

[355] ROSALES, H.M.L., DUGUAY-DROUIN, P., ARCHAMBAULT, L., BEDDAR, S., BEAULIEU, L., Optimization of a multipoint plastic scintillator dosimeter for high dose rate brachytherapy, Med. Phys. **46** (2019) 2412–2421.

[356] SUCHOWERSKA, N., et al., Clinical trials of a urethral dose measurement system in brachytherapy using scintillation detectors, Int. J. Radiat. Oncol. Biol. Phys. **79** (2011) 609–615.

[357] THERRIAULT-PROULX, F., BEAULIEU, L., BEDDAR, S., Validation of plastic scintillation detectors for applications in low-dose-rate brachytherapy, Brachytherapy **16** (2017) 903–909.

[358] RIVARD, M.J., VENSELAAR, J.L., BEAULIEU, L., The evolution of brachytherapy treatment planning, Med. Phys. **36** (2009) 2136–2153.

[359] BEAULIEU, L., et al., Report of the Task Group 186 on model-based dose calculation methods in brachytherapy beyond the TG-43 formalism: current status and recommendations for clinical implementation, Med. Phys. **39** (2012) 6208–6236.

[360] BALLESTER, F., et al., A generic high-dose rate (192)Ir brachytherapy source for evaluation of model-based dose calculations beyond the TG-43 formalism, Med. Phys. **42** (2015) 3048–3061.

[361] SLOBODA, R.S., MORRISON, H., CAWSTON-GRANT, B., MENON, G.V., A brief look at model-based dose calculation principles, practicalities, and promise, J. Contemp. Brachytherapy **9** (2017) 79–88.

[362] RIVARD, M.J., BEAULIEU, L., MOURTADA, F., Enhancements to commissioning techniques and quality assurance of brachytherapy treatment planning systems that use model-based dose calculation algorithms, Med. Phys. **37** (2010) 2645–2658.

[363] BENTLEY, R.E., NATIONAL MEASUREMENT INSTITUTE, Uncertainty in Measurement : the ISO Guide, National Measurement Institute, [Lindfield, N.S.W.] (2005).

[364] INTERNATIONAL ATOMIC ENERGY AGENCY, Calibration of Radiation Protection Monitoring Instruments, Safety Report Series No.16, IAEA, Vienna (2000).

ABBREVIATIONS

AAPM	American Association of Physicists in Medicine
ADCL	accredited dosimetry calibration laboratory
AIST-NMIJ	National Institute of Advanced Industrial Science and Technology – National Metrology Institute of Japan
AKS	air kerma strength
APMP	Asia Pacific Metrology Programme
ARPANSA	Australian Radiation Protection and Nuclear Safety Agency
BARC	Bhabha Atomic Research Centre
BIPM	Bureau International des Poids et Mesures
BRAPHYQS	Brachytherapy Physics Quality Assurance System
BSR	Brachytherapy Source Registry
CIPM	International Committee for Weights and Measures
CMI	Cesky Metrologicky Institut
DIN	Deutsches Institut für Normung
DIRAC	Directory of Radiotherapy Centres
EBRT	external beam radiotherapy
eBT	electronic brachytherapy
ENEA-INMRI	Ente per le Nuove Tecnologie, l'Energia e l'Ambiente – Istituto Nazionale di Metrologia delle Radiazioni Ionizzanti
ESTRO	European Society for Radiotherapy and Oncology
EURAMET	European Association of National Metrology Institutes
GROVEX	Grossvolumen Extrapolationskammer (large-volume parallel-plate extrapolation chamber)
HDR	high dose rate
ICRU	International Commission on Radiation Units and Measurements
IEC	International Electrotechnical Commission
IMS	International Measurement System
INER	Institute of Nuclear Energy Research
IPEM	Institute of Physics and Engineering in Medicine
IROC	Imaging and Radiation Oncology Core
ISO	International Organization for Standardization
IVBT	intravascular brachytherapy
KCDB	Key Comparison Database
KRISS	Korea Research Institute of Standards and Science
LAVV	large-angle variable-volume ionization chamber
LCR	Laboratório de Ciências Radiológicas

LDR	low dose rate
LET	linear energy transfer
LNE-LNHB	Laboratoire National de metrologie et d'Essais - Laboratoire National Henri Becquerel
MDA	minimum detectable activity
NIM	National Institute of Metrology
NIST	National Institute of Standards and Technology
NMI	National Metrology Institute
NPL	National Physical Laboratory
NRC	National Research Council
OSLD	optically stimulated luminescence dosimeter
PDD	percentage depth dose
PDR	pulsed dose rate
PMMA	poly(methyl methacrylate)
PSDL	primary standards dosimetry laboratory
PTB	Physikalisch-Technische Bundesanstalt
QA	quality assurance
QC	quality control
RAKR	reference air kerma rate
RPLD	radiophotoluminescence dosimeter
SSDL	secondary standards dosimetry laboratory
TG	Task Group
TLD	thermoluminescence dosimeter
TPS	treatment planning system
UW	University of Wisconsin
UWADCL	University of Wisconsin Accredited Dosimetry Calibration Laboratory
VAFAC	variable-aperture free air chamber
VNIIM	D. I. Mendeleev All-Russian Institute for Metrology
VSL	Van Swinden Laboratorium
WAFAC	wide-angle free air chamber

CONTRIBUTORS TO DRAFTING AND REVIEW

Almeida, C.E.	Rio de Janeiro State University, Brazil
Bokulic, T.	Faculty of Science, University of Zagreb, Croatia
Carrara, M.	International Atomic Energy Agency
de Pooter, J.	VSL National Metrology Institute, Netherlands
DeWerd, L.A.	University of Wisconsin, United States of America
McEwen, M.	National Research Council, Canada
Rivard, M.J.	Brown University, United States of America
Sander, T.	National Physical Laboratory, United Kingdom
Schneider, T.	Physikalisch-Technische Bundesanstalt, Germany
Toroi, P.	STUK - Radiation and Nuclear Safety Authority, Finland
van der Merwe, D.	International Atomic Energy Agency

Consultants Meetings

Vienna, Austria: 10–14 December 2018

IAEA
International Atomic Energy Agency

ORDERING LOCALLY

IAEA priced publications may be purchased from the sources listed below or from major local booksellers.

Orders for unpriced publications should be made directly to the IAEA. The contact details are given at the end of this list.

NORTH AMERICA

Bernan / Rowman & Littlefield
15250 NBN Way, Blue Ridge Summit, PA 17214, USA
Telephone: +1 800 462 6420 • Fax: +1 800 338 4550
Email: orders@rowman.com • Web site: www.rowman.com/bernan

REST OF WORLD

Please contact your preferred local supplier, or our lead distributor:

Eurospan Group
Gray's Inn House
127 Clerkenwell Road
London EC1R 5DB
United Kingdom

Trade orders and enquiries:
Telephone: +44 (0)176 760 4972 • Fax: +44 (0)176 760 1640
Email: eurospan@turpin-distribution.com

Individual orders:
www.eurospanbookstore.com/iaea

For further information:
Telephone: +44 (0)207 240 0856 • Fax: +44 (0)207 379 0609
Email: info@eurospangroup.com • Web site: www.eurospangroup.com

Orders for both priced and unpriced publications may be addressed directly to:
Marketing and Sales Unit
International Atomic Energy Agency
Vienna International Centre, PO Box 100, 1400 Vienna, Austria
Telephone: +43 1 2600 22529 or 22530 • Fax: +43 1 26007 22529
Email: sales.publications@iaea.org • Web site: www.iaea.org/publications